10 Florida FAST Grade 8 Math Practice Tests

The Ultimate Test Prep Collection with Answer Explanations

Dr. A. Nazari

10 Practice Tests

Welcome!

This book contains **10 full-length practice tests** — the most comprehensive preparation you can get for your Grade 8 math assessment. Each test covers all six topics:

Irrational Numbers   Powers & Scientific Notation

Linear Equations   Functions

Geometry   Data & Relationships

Ten tests give you the practice needed to walk into the real test feeling fully prepared.

Thorough preparation leads to outstanding results.

> **❝** Ten full tests! By the time you finish, there won't be any surprises on test day. **❞**

 How to Use This Book

A complete 10-test preparation program

📋 What's Inside

- **10 Full-Length Practice Tests** — each covers all 6 chapters of Grade 8 math: irrational numbers, exponents & scientific notation, linear equations, functions, geometry, and data analysis.
- **Detailed Answer Explanations** — every question includes a step-by-step solution so you learn from every mistake.
- **Formula Reference Sheet** — all the key Grade 8 formulas you need, organized and ready for quick review.
- **Test Tracker** — log your scores across all 10 tests and monitor your progress from start to finish.

🕐 Your 10-Test Training Plan

⭐ PHASE 1: Foundation (Tests 1–3)

Untimed or soft-timed. Focus on understanding the format, identifying strengths and weaknesses, and building good study habits.

⭐⭐ PHASE 2: Building Skills (Tests 4–7)

Timed (70 minutes each). Work on pacing, accuracy, and showing complete solutions. Review weak topics between tests.

⭐⭐⭐ PHASE 3: Test-Day Ready (Tests 8–10)

Full test conditions: strict timing, quiet space, no notes. Compare scores with your early tests to see your growth.

Schedule: *Take one test every 3–4 days, or one per week. Use study days between tests to review.*

Types of Questions

● **Multiple Choice:** Four options — work the problem first, then match. Eliminate obviously wrong answers to narrow your choices.

✎ **Short Answer & Constructed Response:** Show every step: equations, substitutions, simplifications. Partial credit rewards correct reasoning even if the final answer is off.

📈 **Graphing & Data Analysis:** Plot points, draw lines, interpret graphs. Label axes clearly.

Tip: Ten tests is a full preparation program. Don't rush. The key is what you do between tests — study, review, and understand your mistakes before moving forward.

Find more at
ViewMath.com/FL-Grade8

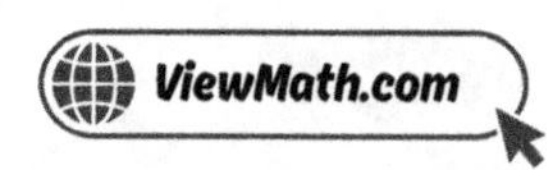

Test-Taking Tips

Your complete test-day toolkit

Before the Test

- Review your notes from the previous test — focus on your weak topics
- Set up a quiet, clean workspace with all your materials ready
- Start with a positive mindset: you've prepared for this

During the Test

- Read each problem fully before calculating anything
- Write the formula or set up the equation first, then substitute values
- Show all your work — every step, every operation
- If stuck for more than 2 minutes, mark it and move on
- Use estimation to check if your answers are reasonable

After the Test

- Read the full explanation for every question you got wrong
- Write down which topics gave you trouble (not just question numbers)
- Study those topics before taking the next test
- Record your score in the Test Tracker

Common Mistakes in Grade 8 Math

⚠ **Exponents:** $(ab)^n = a^n b^n$, but $a^m + a^n \neq a^{m+n}$. Only multiply/divide to combine.

⚠ **Slope formula:** $m = \frac{y_2 - y_1}{x_2 - x_1}$ — keep the order consistent.

⚠ **Systems of equations:** The solution must satisfy both equations.

⚠ **Transformations:** Rotations and reflections change position; dilations change size.

⚠ **Volume:** Use $\pi \approx 3.14$ or leave as π — match what the question asks.

Find more at
ViewMath.com/FL-Grade8

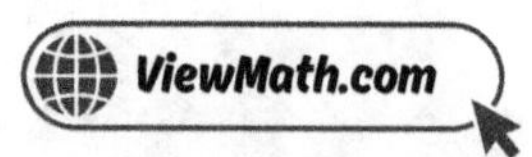

What You'll Need

Gather these materials before you begin

Sharpened Pencils — #2 pencils, at least two

Good Eraser — for clean corrections

Scratch Paper — for working out problems

Ruler / Straightedge — for graphing & geometry

Quiet Space — no distractions

Focused Mind — ready to do your best

Allowed Materials

- ✔ Pencils and eraser
- ✔ Scratch paper (provided on official test day)
- ✔ Ruler or straightedge (if required)
- ✔ Protractor (if required)

Not Allowed

- ✖ Calculator (unless your state test allows it)
- ✖ Cell phone or any electronic device
- ✖ Notes, textbooks, or reference sheets
- ✖ Help from others during the test

Ten tests is a comprehensive program. Plan **one test every 3–4 days** (or one per week) with study sessions between each test.

How to help:

- Tests 1–3 should be untimed — build understanding before adding pressure.
- After each test, review the answer explanations together. Ask: "Which topics were hardest? Let's study those before the next one."
- Use the Test Tracker to celebrate progress over time.
- For topic-specific help, pair this book with our **Grade 8 Math Study Guide** or **Grade 8 Workbook**.

Grade 8 Formula Reference

Keep this page handy — you may use it during your practice tests!

X^1 Exponent Rules

$$a^m \cdot a^n = a^{m+n} \qquad (a^m)^n = a^{mn} \qquad (ab)^n = a^n \cdot b^n$$

$$\frac{a^m}{a^n} = a^{m-n} \qquad a^0 = 1 \ (a \neq 0) \qquad a^{-n} = \frac{1}{a^n}$$

Lines & Linear Equations

Slope: $m = \dfrac{y_2 - y_1}{x_2 - x_1} = \dfrac{rise}{run}$ **Slope-intercept:** $y = mx + b$ **Proportional:** $y = mx$

m = slope b = y-intercept   *Parallel lines: same slope*   *Proportional: passes through origin*

Scientific Notation

$a \times 10^n$ where $1 \leq |a| < 10$ **Multiply:** add exponents   **Divide:** subtract exponents

$\sqrt{x}$ Roots & Number Sense

Perfect squares: 1, 4, 9, 16, 25, 36, 49, 64, 81, 100, 121, 144

Perfect cubes: 1, 8, 27, 64, 125 $\sqrt{2} \approx 1.414$ $\sqrt{3} \approx 1.732$ $\pi \approx 3.14159$

Pythagorean Theorem & Distance

$a^2 + b^2 = c^2$ c = hypotenuse (longest side of a right triangle)   **Distance:** $d = \sqrt{(x_2 - x_1)^2 + (y_2 - y_1)^2}$

Volume Formulas

Cylinder $V = \pi r^2 h$ **Cone** $V = \dfrac{1}{3}\pi r^2 h$ **Sphere** $V = \dfrac{4}{3}\pi r^3$

⌘ Angle Relationships

Triangle angle sum: $180°$ **Exterior angle** = sum of two remote interior angles

Parallel lines + transversal: Alternate interior angles are equal · Co-interior angles sum to $180°$

⌁ Functions

Each input → exactly one output   **Vertical line test:** if any vertical line hits graph more than once ⇒ not a function

Linear: constant rate of change $(y = mx + b)$ **Nonlinear:** rate of change varies

↻ Transformations

Translation: slide   **Reflection:** flip   **Rotation:** turn   **Dilation:** resize

Congruent = same shape & size   Similar = same shape, proportional size

Tip: Bookmark this page! Review it before each test so these formulas become second nature.

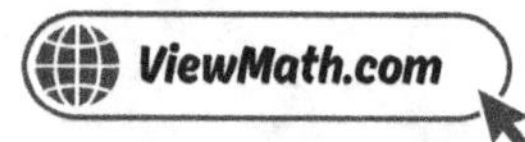

▦ Multiplication Table ▦

You may use this table during your practice tests!

×	1	2	3	4	5	6	7	8	9	10	11	12
1	1	2	3	4	5	6	7	8	9	10	11	12
2	2	4	6	8	10	12	14	16	18	20	22	24
3	3	6	9	12	15	18	21	24	27	30	33	36
4	4	8	12	16	20	24	28	32	36	40	44	48
5	5	10	15	20	25	30	35	40	45	50	55	60
6	6	12	18	24	30	36	42	48	54	60	66	72
7	7	14	21	28	35	42	49	56	63	70	77	84
8	8	16	24	32	40	48	56	64	72	80	88	96
9	9	18	27	36	45	54	63	72	81	90	99	108
10	10	20	30	40	50	60	70	80	90	100	110	120
11	11	22	33	44	55	66	77	88	99	110	121	132
12	12	24	36	48	60	72	84	96	108	120	132	144

💡 How to Use This Table

To find **4 × 7**:

1. Find **4** in the left column (blue).
2. Find **7** in the top row (blue).
3. Follow the row and column until they meet: the answer is **28**!

Tip: You can also use this table for division! If you know $28 \div 4 = $?, find 28 in the 4's row. The column header gives you the answer: **7**!

My Test Tracker

Record every test and watch your scores improve

Name: ___________________________   **Start Date:** ___________________

GETTING STARTED (Tests 1–3)

Test 1 — Untimed

Date: ______________   Score: ______ / ______   %: ______   Topics to review: ______________________

Test 2 — Untimed

Date: ______________   Score: ______ / ______   %: ______   Topics to review: ______________________

Test 3 — Soft Timer

Date: ______________   Score: ______ / ______   %: ______   Topics to review: ______________________

BUILDING SKILLS (Tests 4–7)

Test 4 — Timed (70 min)

Date: ______________   Score: ______ / ______   %: ______   Focus area: ______________________

Test 5 — Timed (70 min)

Date: ______________   Score: ______ / ______   %: ______   Focus area: ______________________

Test 6 — Timed (70 min)

Date: ______________   Score: ______ / ______   %: ______   Focus area: ______________________

Test 7 — Timed (70 min)

Date: ______________   Score: ______ / ______   %: ______   Focus area: ______________________

TEST-DAY READY (Tests 8–10)

Date: _____________   Score: _______ / _______   %: _______   Growth since Test 1: _____________________

Date: _____________   Score: _______ / _______   %: _______   Growth since Test 1: _____________________

Date: _____________   Score: _______ / _______   %: _______   Growth since Test 1: _____________________

📊 Score Progress

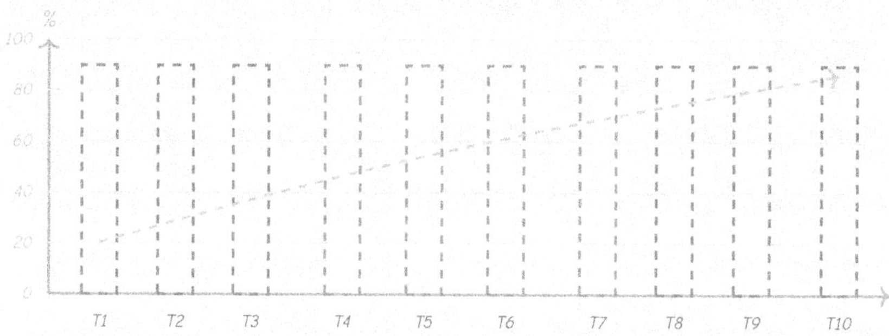

Shade each bar after every test. Watch your improvement!

✅ Final Reflection

The most important thing I learned: ____

The topic where I improved the most: _______________________________________

My advice for other students: ____

 Let's learn and have fun!

Practice Test 1

30 Questions

✏️ Before You Start ✏️

- ✓ **Read each question carefully** before choosing your answer.
- ✓ **Show your work** on scratch paper when you need to.
- ✓ **Skip hard questions** and come back to them later.
- ✓ **Check your answers** when you're done.
- ✓ **Take your time** — there's no rush!

⭐ You've Got This! ⭐

Do your best and show what you know!

1. True or false: The number $\frac{22}{7}$ is equal to π.

   Your Answer

2. The number line below shows $\sqrt{n}$ placed between two tick marks. Based on the position, what could n be?

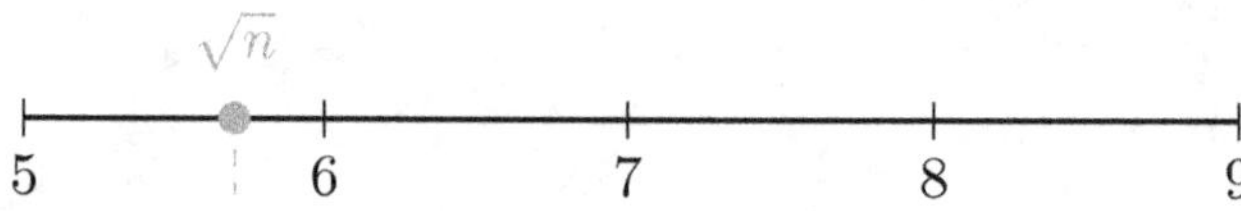

   (A) $n = 28$

   (B) $n = 33$

   (C) $n = 40$

   (D) $n = 55$

3. Look at the receipt below. The sales tax rate is 8%. What should the total be?

Item	Price
T-shirt	$20.00
Jeans	$45.00
Subtotal	$65.00
Tax (8%)	?
Total	?

   (A) Tax = $5.00; Total = $70.00

   (B) Tax = $5.20; Total = $70.20

   (C) Tax = $6.50; Total = $71.50

   (D) Tax = $8.00; Total = $73.00

4. Simplify $7^5 \cdot 7^{-5}$.

   (A) 7^{25}

   (B) 7^{10}

   (C) 0

   (D) 1

5. The diagram shows a square and a cube. The square has an area of A square units and the cube has a volume of V cubic units. If $A = V$, which pair of side lengths is correct?

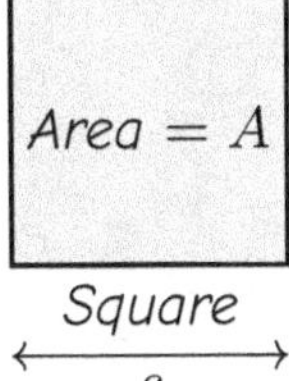

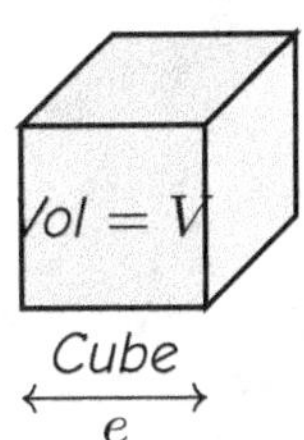

(A) $s = 6,\ e = 6$

(B) $s = 8,\ e = 4$

(C) $s = 9,\ e = 3$

(D) $s = 5,\ e = 5$

6. Earth's mass is about 6×10^{24} kg and Jupiter's mass is about 1.9×10^{27} kg. About how many times more massive is Jupiter than Earth? Round to the nearest whole number.

Your Answer:

7. The table and the equation below each describe a proportional relationship.

x	1	2	3	4
y	6	12	18	24

Equation: $y = 5x$

Which relationship has the larger constant of proportionality?

(A) The table

(B) The equation

(C) They are equal.

(D) Not enough information.

8. A phone plan charges $0.15 per text. If you graph total cost vs. number of texts, the slope is:

(A) 15

(B) 0.15

(C) 1.50

(D) 0.015

Find more at
ViewMath.com/FL-Grade8

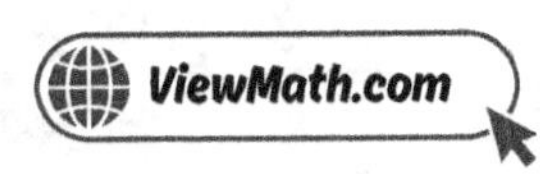

9. *Solve* $-5x + 14 = -x + 2$.

Your Answer:

10. A school play sold 200 tickets. Adult tickets were \$8 and student tickets were \$5. Total revenue was \$1,240. How many adult tickets were sold?

Your Answer:

11. Use the graph of g below. For which input is $g(x) = 0$?

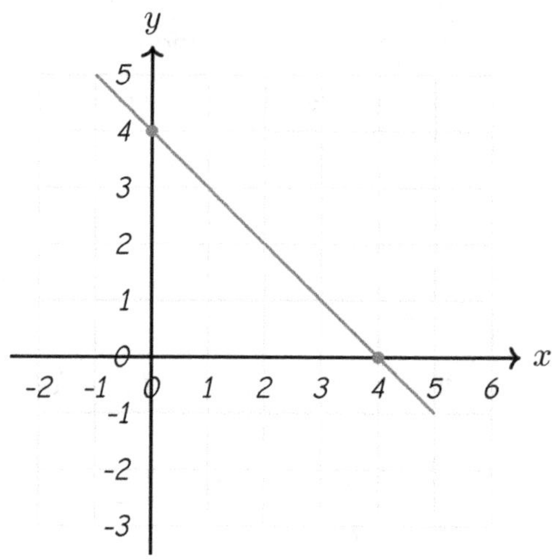

(A) $x = 0$

(B) $x = 2$

(C) $x = 4$

(D) $x = -4$

12. The tables below show two functions. Find the rate of change and initial value of each. Which function will have a greater value at $x = 10$?

Function A

x	y
0	20
1	23
2	26
3	29

Function B

x	y
0	5
1	10
2	15
3	20

Your Answer:

13. Does the equation $y = -12$ represent a linear function? Write Yes or No.

Your Answer:

14. A taxi charges a \$4 flat fee plus \$3 per mile. Which equation models the total cost y for x miles?

(A) $y = 4x + 3$

(B) $y = 3x + 4$

(C) $y = 7x$

(D) $y = 3x - 4$

15. A graph shows a line going upward with a constant slope. Is the rate of change constant or varying?

Your Answer:

16. Which of the following is NOT preserved under a translation?

(A) Side lengths

(B) Angle measures

(C) Position

(D) Parallelism of sides

17. Describe a sequence of rigid transformations that maps a triangle with vertices $(1,1)$, $(4,1)$, $(1,3)$ to a triangle with vertices $(-1,1)$, $(-4,1)$, $(-1,3)$.

Your Answer:

18. Point $(5,2)$ is rotated $180°$ around the origin. What are the coordinates of the image?

Your Answer:

19. A dilation centered at the origin maps $(4,-2)$ to $(12,-6)$. What is the scale factor?

(A) 2

(B) 3

(C) 4

(D) 6

20. Parallel lines ℓ and m are cut by transversal t. If a pair of alternate interior angles are $(4x)°$ and $80°$, what is x?

(A) 320

(B) 20

(C) 25

(D) 10

21. A right triangle has legs 5 and 5. What is the hypotenuse?

(A) 10

(B) $\sqrt{50}$

(C) 25

(D) $\sqrt{10}$

22. What is the distance between $(2,-5)$ and $(6,-2)$?

(A) 25

(B) 7

(C) 5

(D) $\sqrt{7}$

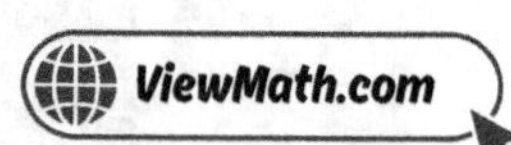

23. A cylinder has a volume of 450π cm³. A cone has the same base and height. What is the cone's volume?

(A) 150π cm³

(B) 225π cm³

(C) 1350π cm³

(D) 900π cm³

24. The complement of an angle is 17°. What is the original angle?

(A) 163°

(B) 73°

(C) 83°

(D) 17°

25. A quadrilateral has angles 85°, 110°, and 75°. What is the fourth angle?

(A) 80°

(B) 90°

(C) 100°

(D) 70°

26. What type of association does this scatter plot show?

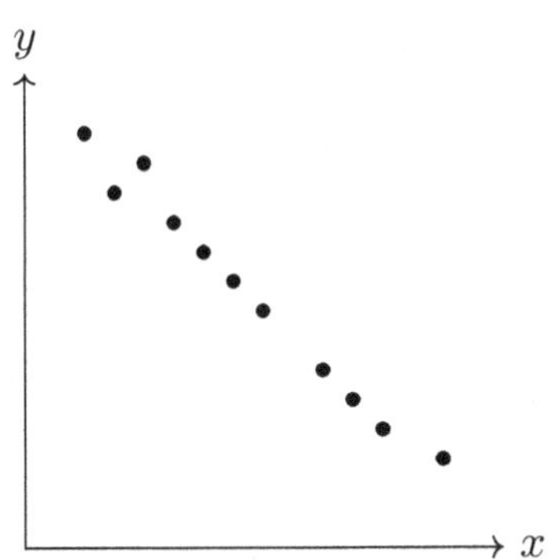

(A) Positive linear

(B) Negative linear

(C) No association

(D) Nonlinear

27. A scatter plot has 10 data points. One outlier is far above the trend. When drawing the line of best fit, you should:

(A) draw the line through the outlier

(B) ignore the outlier and fit the line to the other 9 points

(C) not draw a line at all

(D) draw the line only through the outlier

28. A trend line gives $y = -3x + 45$. Data was collected for $x = 1$ to $x = 10$. Predict y when $x = 5$ and when $x = 30$. State which prediction is more reliable and why.

Your Answer:

29. Using the table above, what percentage of 7th graders ride a bike?

(A) 55%

(B) 45%

(C) 62.5%

(D) 37.5%

30. A deck has 52 cards. Two cards are drawn without replacement. Find $P(\text{both aces})$.

Your Answer:

 # End of Practice Test 1 

Great job finishing the test!

☑ My Score

I got __________ out of 30 questions right.

Check your answers in the **Answer Key** at the back of the book.

💡 Review any questions you missed. That's how we learn!

📊 Check Your Score Online!

Visit **ViewMath Academy** to enter your answers and see which topics you need to review. You can also explore lessons, take quizzes, track your scores, and save your progress!

viewmath.com/score/8.1.FL.16

Or go to *viewmath.com/score* and enter code: 8.1.FL.16

2

Practice Test 2

 30 Questions

---

- ✔ **Read each question carefully** before choosing your answer.
- ✔ **Show your work** on scratch paper when you need to.
- ✔ **Skip hard questions** and come back to them later.
- ✔ **Check your answers** when you're done.
- ✔ **Take your time** — there's no rush!

★ You've Got This! ★

Do your best and show what you know!

1. The number line below shows two points, P and Q.

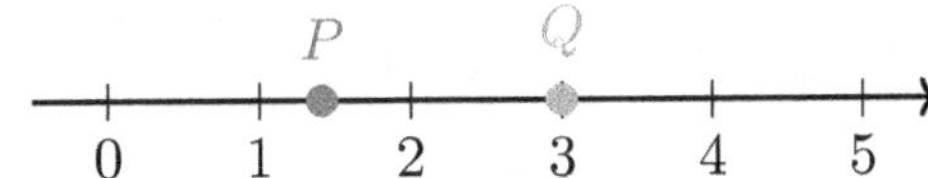

Point P represents $\sqrt{2}$ and point Q represents 3. Which statement is true?

(A) Both P and Q are rational.

(B) Both P and Q are irrational.

(C) P is irrational and Q is rational.

(D) P is rational and Q is irrational.

2. On the number line below, each tick mark represents 0.1 units. Plot the approximate location of $\sqrt{5}$ and label it.

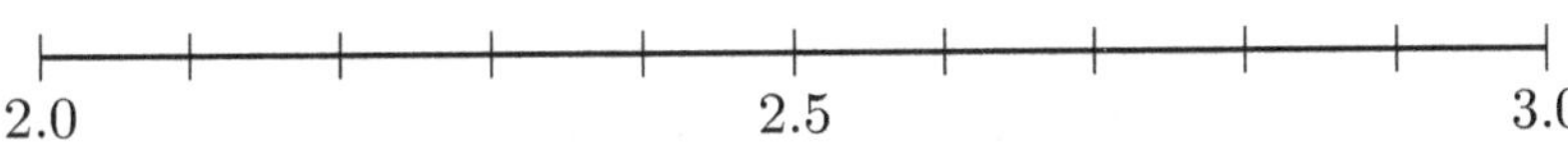

Your Answer

3. The price tag below shows a $90 item with a 20% discount. After the discount, 6% sales tax is added. Find the final price.

Your Answer:

4. Which expression is equivalent to 4^{-3}?

   (A) -64

   (B) -12

   (C) $\frac{1}{12}$

   (D) $\frac{1}{64}$

5. Solve $x^2 = \frac{9}{16}$.

   (A) $x = \frac{3}{4}$ only

   (B) $x = \pm\frac{3}{4}$

   (C) $x = \frac{9}{8}$

   (D) $x = \pm\frac{9}{4}$

6. What is $9.1 \times 10^5 - 8.6 \times 10^5$?

   (A) 0.5×10^5

   (B) 5×10^4

   (C) 5×10^5

   (D) Both A and B

7. A graph of a proportional relationship passes through the point $(5, 20)$. What is the constant of proportionality k?

   (A) 5

   (B) 15

   (C) 4

   (D) 100

8. A jogger runs 2 miles in 20 minutes and 5 miles in 50 minutes. What is the slope of distance vs. time (in miles per minute)?

   Your Answer

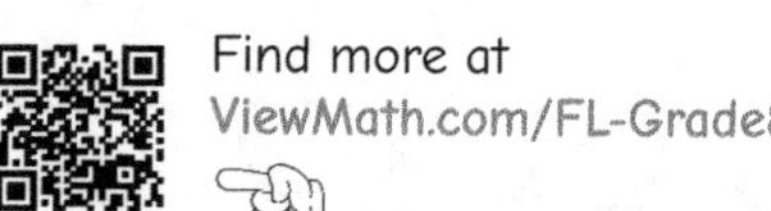

Find more at
ViewMath.com/FL-Grade8

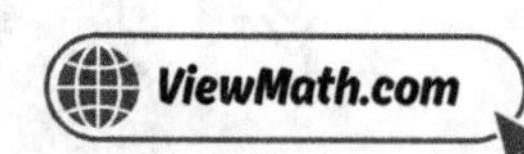

9. Solve $\frac{x}{4} + 3 = 7$.

   (A) $x = 1$ (B) $x = 10$

   (C) $x = 16$ (D) $x = 28$

10. You have \$5 bills and \$10 bills totaling \$85. You have 12 bills. How many \$10 bills do you have?

Your Answer:

11. The function $f(x) = 10 - 3x$ gives the number of pieces of candy remaining after x friends each take 3. How many pieces are left after 2 friends take candy?

   (A) 1   (B) 4

   (C) 7   (D) 16

12. A table for Function C shows $(0, 7)$, $(1, 11)$, $(2, 15)$. What is the initial value?

Your Answer:

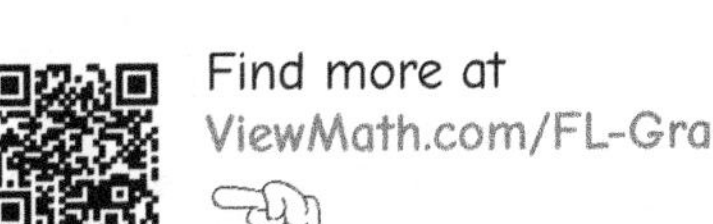

13. Look at the graph below. Is the function linear or nonlinear? Explain your reasoning.

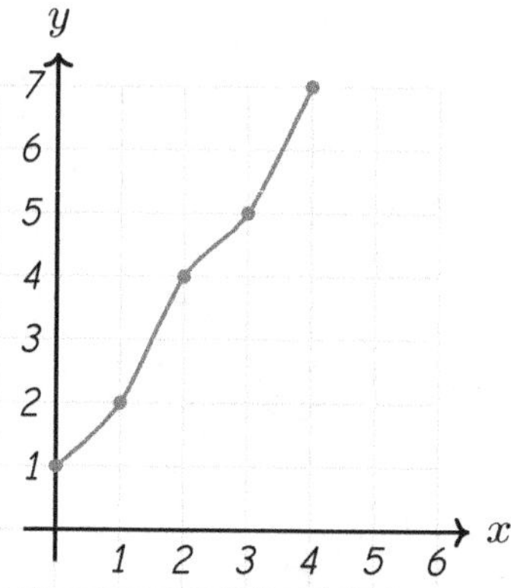

Your Answer:

14. The table shows $(2, 9)$ and $(6, 25)$. Find the slope of the line.

Your Answer:

15. A graph is a straight line segment going upward. The function during this segment is:

(A) Nonlinear and increasing

(B) Linear and decreasing

(C) Linear and increasing

(D) Nonlinear and constant

16. A figure is reflected over the y-axis. What happens to point $(7, -2)$?

(A) $(7, 2)$

(B) $(-7, 2)$

(C) $(-7, -2)$

(D) $(-2, 7)$

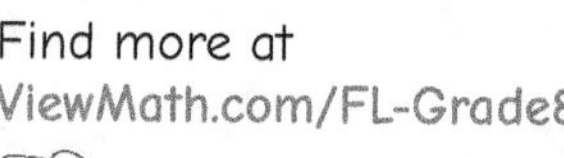
Find more at
ViewMath.com/FL-Grade8

17. *Triangle P has angles $40°$, $60°$, and $80°$. Triangle Q has angles $40°$, $60°$, and $80°$. Are they congruent?*

(A) *Yes, because matching angles guarantee congruence.*

(B) *No, because angles alone do not guarantee congruence.*

(C) *Yes, because all three angles match.*

(D) *No, because the angles are in different order.*

18. *A point is translated by $(3, -2)$ and then reflected over the y-axis. If the original point is $(1, 4)$, what is the final image?*

(A) $(-4, 2)$

(B) $(4, 2)$

(C) $(-4, -2)$

(D) $(4, -2)$

19. *Two similar rectangles have a scale factor of 3. The smaller has area 12 cm^2. What is the area of the larger?*

Your Answer

20. *Two parallel lines are cut by a transversal. One angle measures $72°$. What is its alternate interior angle?*

(A) $18°$

(B) $72°$

(C) $108°$

(D) $288°$

21. *A right triangle has legs 8 and 15. What is the hypotenuse?*

(A) 23

(B) 17

(C) $\sqrt{23}$

(D) 289

22. *Find the distance between $(5, -2)$ and $(5, 7)$.*

Your Answer

Find more at
ViewMath.com/FL-Grade8

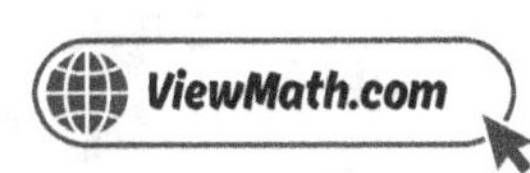

23. What is the volume of a cylinder with radius 7 m and height 3 m? Leave your answer in terms of π.

(A) $21\pi\ m^3$

(B) $63\pi\ m^3$

(C) $147\pi\ m^3$

(D) $49\pi\ m^3$

24. True or false: Vertical angles are always supplementary.

Your Answer:

25. What is the value of x?

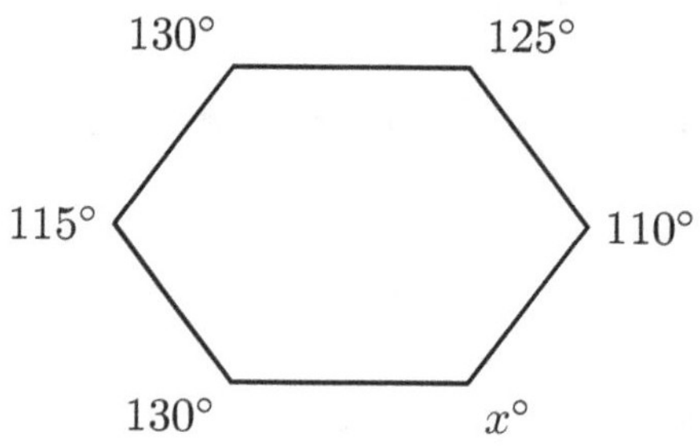

(A) 100°

(B) 110°

(C) 120°

(D) 110°

26. Describe the association shown in this scatter plot. Mention direction, shape, and any unusual features.

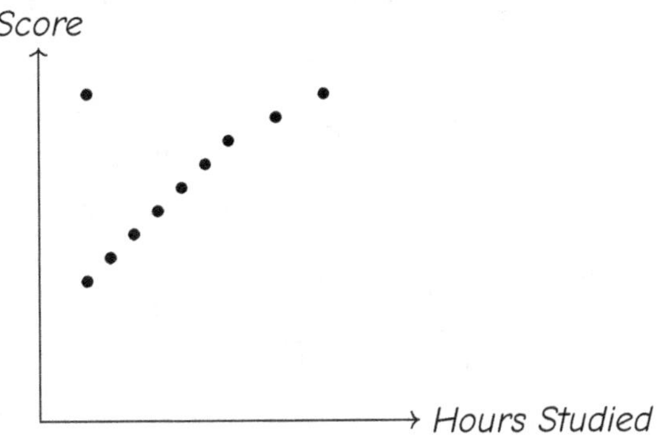

Your Answer

27. What is a line of best fit?

(A) A line that passes through every data point

(B) A line that connects the first and last data points

(C) A straight line that comes close to most of the data points

(D) A vertical line through the middle of the scatter plot

28. Which statement about slope in a linear model is true?

(A) A positive slope means y decreases as x increases.

(B) A negative slope means y increases as x increases.

(C) A slope of zero means y stays constant.

(D) The slope is always positive in real data.

29. *A segmented (stacked) bar chart shows the proportion of students choosing band or chorus by gender. The boys' bar is 60% band and 40% chorus. The girls' bar is 45% band and 55% chorus. What can you conclude?*

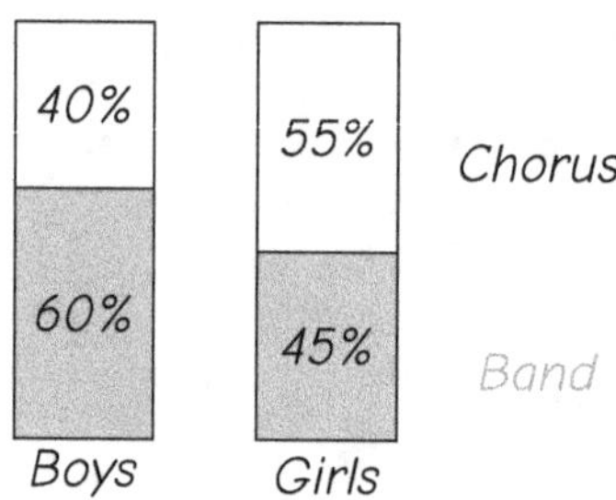

(A) There is no association between gender and music choice.

(B) Boys are more likely to choose band; girls more likely to choose chorus.

(C) Girls and boys prefer band equally.

(D) Chorus is more popular overall.

30. *A tree diagram shows flipping a coin and rolling a die (1–6). How many total outcomes are in the sample space?*

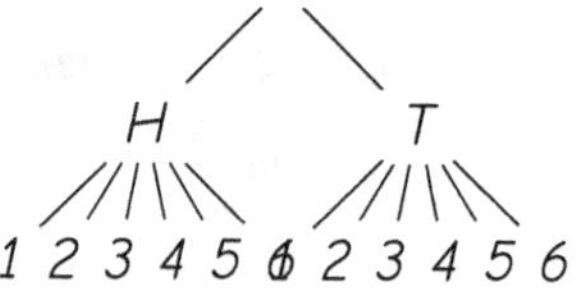

(A) 6

(B) 8

(C) 12

(D) 36

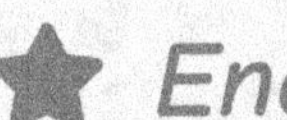 # End of Practice Test 2 

Great job finishing the test!

✅ My Score

I got _____________ out of 30 questions right.

*Check your answers in the **Answer Key** at the back of the book.*

💡 *Review any questions you missed. That's how we learn!*

📊 Check Your Score Online!

Visit **ViewMath Academy** to enter your answers and see which topics you need to review. You can also explore lessons, take quizzes, track your scores, and save your progress!

viewmath.com/score/8.1.FL.17

Or go to viewmath.com/score and enter code: 8.1.FL.17

3

Practice Test 3

 30 Questions

✏ Before You Start ✏

- ✓ **Read each question carefully** before choosing your answer.
- ✓ **Show your work** on scratch paper when you need to.
- ✓ **Skip hard questions** and come back to them later.
- ✓ **Check your answers** when you're done.
- ✓ **Take your time** — there's no rush!

⭐ **You've Got This!** ⭐

 Do your best and show what you know!

1. Which number below can be written as a fraction $\frac{a}{b}$ where a and b are integers and $b \neq 0$?

(A) $\sqrt{3}$

(B) π

(C) $0.\overline{81}$

(D) $\sqrt{11}$

2. Which is the best approximation of $\sqrt{30}$ to one decimal place?

(A) 5.1

(B) 5.3

(C) 5.5

(D) 5.8

3. Jake borrows $2,000 at 6% simple interest for 2 years. How much total does he pay back?

(A) $2,120

(B) $2,240

(C) $2,600

(D) $2,012

4. Which of the following is equal to $\frac{8^6}{8^6}$?

(A) 8^{36}

(B) 8^0

(C) 0

(D) 8

5. A square has an area of 196 square centimeters. What is the length of one side?

(A) 13 cm

(B) 49 cm

(C) 14 cm

(D) 98 cm

6. What is $\frac{8\times10^7}{2\times10^3}$?

(A) 4×10^4

(B) 6×10^4

(C) 4×10^{10}

(D) 16×10^{10}

7. Machine A fills $y=12x$ bottles per hour. Machine B fills 50 bottles in 5 hours. Which machine is faster?

(A) Machine A

(B) Machine B

(C) They fill at the same rate.

(D) Cannot be determined.

8. A line passes through $(2,7)$ and $(5,1)$. What is the slope?

(A) -2

(B) 2

(C) -3

(D) 3

9. Solve $0.5x+1.5=4$.

(A) $x=3$

(B) $x=5$

(C) $x=7$

(D) $x=11$

10. Two trains leave the same station heading in opposite directions. Train A goes 60 mph and Train B goes 80 mph. After how many hours are they 420 miles apart?

(A) 2

(B) 2.5

(C) 3

(D) 3.5

Find more at
ViewMath.com/FL-Grade8

11. A function is defined by the table below.

x	0	1	2	3
$f(x)$	-1	3	7	11

What is $f(0) + f(2)$?

(A) 2

(B) 6

(C) 8

(D) 10

12. Function A: $y = -x + 10$. Function B: $y = -4x + 10$. Which function decreases faster? Write A or B.

Your Answer:

13. A table shows $(0, 3)$, $(1, 7)$, $(2, 11)$, $(3, 15)$. Is this linear or nonlinear?

(A) Nonlinear, because the outputs are odd numbers only.

(B) Linear, because the change in y is always 4.

(C) Nonlinear, because the outputs get larger.

(D) Linear, because the first output is positive.

14. A line passes through $(0, 4)$ and $(5, 24)$. Write its equation.

Your Answer:

15. A graph of water in a tank is a straight line sloping downward. What is happening?

(A) Water is being added at a constant rate.

(B) Water is leaking out at a constant rate.

(C) The tank is full and not changing.

(D) Water is being added at an increasing rate.

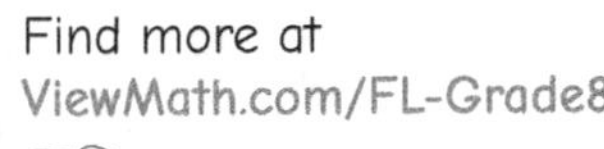

Find more at
ViewMath.com/FL-Grade8

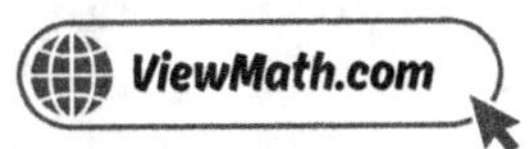

16. Point $R(-1, 4)$ is rotated 90° counterclockwise around the origin. What are the coordinates of R'?

(A) $(4, 1)$

(B) $(-4, -1)$

(C) $(1, -4)$

(D) $(-4, 1)$

17. Two quadrilaterals are shown below. Which statement is correct?

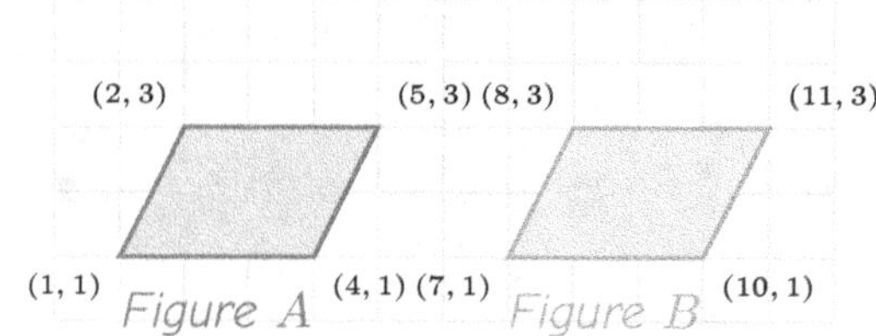

(A) They are congruent — mapped by a translation of 6 units right.

(B) They are not congruent — different side lengths.

(C) They are similar but not congruent.

(D) They are congruent — mapped by a reflection.

18. Point $A(0, -5)$ is rotated 90° counterclockwise around the origin. Where does A' land?

(A) $(5, 0)$

(B) $(0, 5)$

(C) $(-5, 0)$

(D) $(0, -5)$

19. Two similar triangles have a scale factor of 4 from the smaller to the larger. If a side of the smaller triangle is 3 cm, what is the corresponding side of the larger triangle?

(A) 7 cm

(B) 0.75 cm

(C) 12 cm

(D) 9 cm

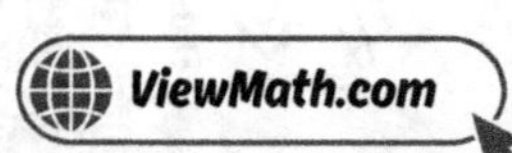

20. In an isosceles triangle, the two base angles are each 54°. What is the vertex angle?

(A) 54°

(B) 72°

(C) 108°

(D) 126°

21. A baseball diamond is a square with 90 ft sides. How far is it from home plate to second base (the diagonal)?

(A) 180 ft

(B) $90\sqrt{2}$ ft

(C) 45 ft

(D) $\sqrt{90}$ ft

22. Find the distance between points A and B shown on the coordinate plane.

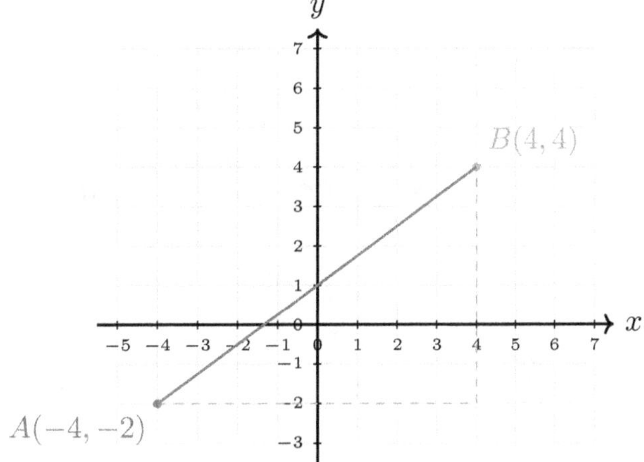

Your Answer:

23. The volume of a sphere with radius r is $\frac{4}{3}\pi r^3$. If the radius is tripled, the new volume is:

(A) 3 times the original

(B) 9 times the original

(C) 27 times the original

(D) 81 times the original

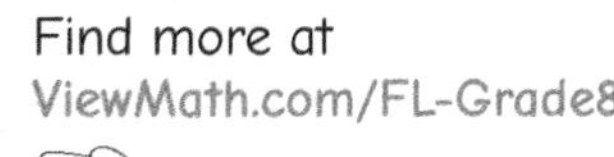
Find more at
ViewMath.com/FL-Grade8
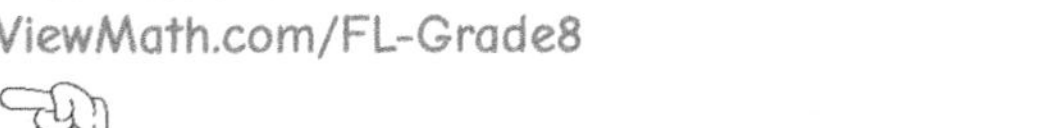

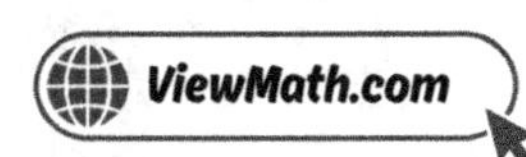

24. *An angle measures 90°. Its supplement measures:*

(A) 0°

(B) 90°

(C) 180°

(D) 270°

25. *As the number of sides of a regular polygon increases, each interior angle:*

(A) decreases toward 0°.

(B) stays the same.

(C) increases toward 180°.

(D) increases toward 360°.

26. *Can a scatter plot show a positive association that is nonlinear? Give an example.*

Your Answer

27. *A scatter plot shows a strong nonlinear (curved) pattern. Should you draw a straight line of best fit?*

(A) Yes — always draw a straight line.

(B) No — a straight line does not fit curved data well.

(C) Yes — but only if there are outliers.

(D) No — you should never draw any line.

28. *Using the model $y = -4x + 100$, predict y when $x = 10$.*

(A) 140

(B) 60

(C) 96

(D) 40

Find more at
ViewMath.com/FL-Grade8

29.

	Dog	Cat	Total
Boys	15	10	25
Girls	12	13	25
Total	27	23	50

How many girls prefer dogs?

(A) 10

(B) 12

(C) 13

(D) 15

30. A fair die is rolled twice. What is $P(\text{both rolls are } 6)$?

(A) $\frac{1}{6}$

(B) $\frac{1}{12}$

(C) $\frac{1}{36}$

(D) $\frac{2}{6}$

 # End of Practice Test 3 

Great job finishing the test!

 My Score

I got _____________ out of 30 questions right.

*Check your answers in the **Answer Key** at the back of the book.*

💡 *Review any questions you missed. That's how we learn!*

📊 Check Your Score Online!

Visit **ViewMath Academy** to enter your answers and see which topics you need to review. You can also explore lessons, take quizzes, track your scores, and save your progress!

viewmath.com/score/8.1.FL.18

Or go to viewmath.com/score and enter code: 8.1.FL.18

Practice Test 4

30 Questions

✏️ Before You Start ✏️

- ✔ **Read each question carefully** before choosing your answer.
- ✔ **Show your work** on scratch paper when you need to.
- ✔ **Skip hard questions** and come back to them later.
- ✔ **Check your answers** when you're done.
- ✔ **Take your time** — there's no rush!

⭐ You've Got This! ⭐

Do your best and show what you know!

1. Which of the following numbers is irrational?

   (A) $\frac{5}{8}$

   (B) $0.\overline{6}$

   (C) $\sqrt{9}$

   (D) $\sqrt{7}$

2. Approximate $\sqrt{68}$ to one decimal place.

   Your Answer:

3. An investment of \$800 earns \$96 in simple interest over 4 years. What is the annual interest rate?

   (A) 2%

   (B) 3%

   (C) 4%

   (D) 12%

4. Evaluate $5^0 + 5^{-1}$. Express your answer as a decimal.

   Your Answer:

5. What is $\sqrt{49}$?

   (A) 6

   (B) 7

   (C) 8

   (D) 24.5

6. What is $5.2 \times 10^6 + 3.8 \times 10^6$?

   (A) 9×10^6

   (B) 9×10^{12}

   (C) 19.76×10^6

   (D) 9×10^{36}

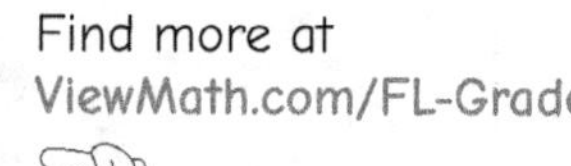

Find more at
ViewMath.com/FL-Grade8

7. *Which ordered pair could NOT lie on the graph of a proportional relationship?*

(A) $(0,0)$

(B) $(1,5)$

(C) $(3,15)$

(D) $(2,13)$

8. *A line passes through $(0,4)$ and $(8,4)$. What is its slope?*

Your Answer:

9. *Solve $\frac{2x-1}{3} = 5$.*

(A) $x = 7$

(B) $x = 7.5$

(C) $x = 8$

(D) $x = 16$

10. *Two numbers add to 25 and differ by 7. What is the larger number?*

(A) 14

(B) 15

(C) 16

(D) 17

11. *If $h(x) = 2x^2$, what is $h(5)$?*

(A) 20

(B) 25

(C) 50

(D) 100

12. *Function A: $y = 2x + 15$. Function B: $y = 5x + 6$. What is the value of each function at $x = 3$?*

Your Answer:

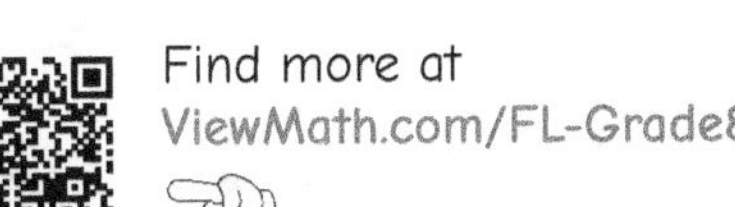
Find more at
ViewMath.com/FL-Grade8

13. *The table below shows a function. Is it linear or nonlinear?*

x	0	1	2	3	4
y	1	2	5	10	17

(A) *Linear, because y always increases.*

(B) *Linear, because x increases by 1 each time.*

(C) *Nonlinear, because the differences in y are $1, 3, 5, 7$ — not constant.*

(D) *Nonlinear, because the outputs are all positive.*

14. *Find the equation of the line through $(1, 3)$ and $(4, 12)$.*

Your Answer:

15. *A bathtub fills up, then someone soaks, then the water drains. Which describes the water level graph?*

(A) *Decreasing, constant, increasing*

(B) *Increasing, decreasing, constant*

(C) *Increasing, constant, decreasing*

(D) *Constant, increasing, decreasing*

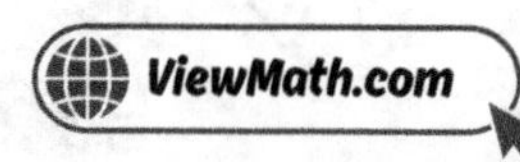

16. *Look at the coordinate plane. Which transformation maps triangle PQR to triangle $P'Q'R'$?*

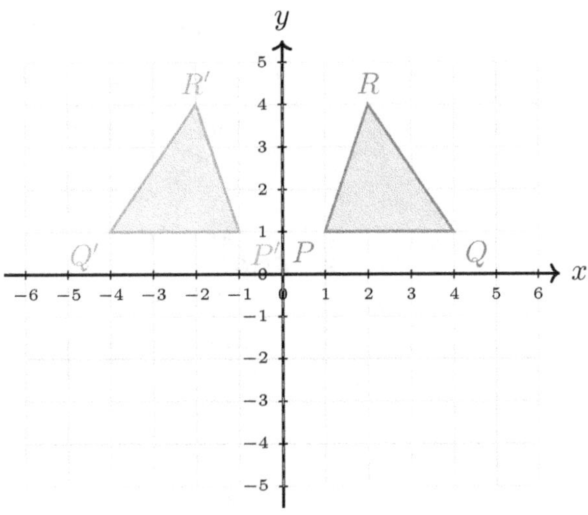

(A) Translation 5 units left

(B) Reflection over the y-axis

(C) Rotation 90° clockwise

(D) Reflection over the x-axis

17. *Rectangle $ABCD$ has $AB = 6$ and $BC = 10$. Rectangle $WXYZ$ has $WX = 10$ and $XY = 6$. Are they congruent?*

(A) No, because the side lengths are in different order.

(B) No, because $6 \neq 10$.

(C) Yes, because the sets of side lengths are the same.

(D) Cannot be determined.

18. *A rectangle has vertex $(-6, 2)$. After a translation of $(4, -5)$, what is the new position of this vertex?*

(A) $(-10, 7)$

(B) $(-2, -3)$

(C) $(2, -3)$

(D) $(-2, 3)$

19. If two figures are congruent, are they also similar?

    (A) Yes, with scale factor $k = 1$.

    (B) No, congruent and similar are different properties.

    (C) Only if they are triangles.

    (D) Only if they have the same orientation.

20. An exterior angle of a triangle is $140°$. What is the adjacent interior angle?

    (A) $40°$

    (B) $140°$

    (C) $50°$

    (D) $220°$

21. Is a triangle with sides 7, 24, 25 a right triangle?

    (A) Yes, because $7 + 24 > 25$.

    (B) Yes, because $7^2 + 24^2 = 25^2$.

    (C) No, because $7^2 + 24^2 \neq 25^2$.

    (D) No, because all sides must be equal.

22. What is the distance between $(-1, 2)$ and $(2, 6)$?

    (A) 25

    (B) 7

    (C) 5

    (D) $\sqrt{13}$

Find more at
ViewMath.com/FL-Grade8

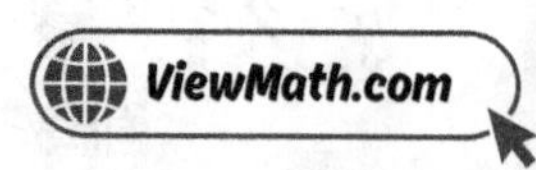

23. What is the volume of the cone shown below in terms of π?

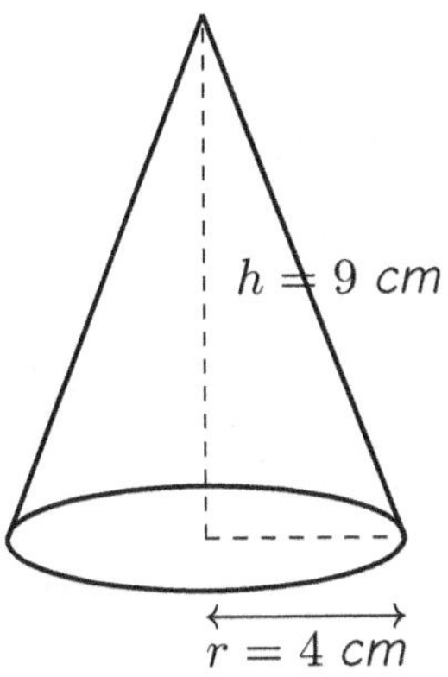

(A) 144π cm^3

(B) 48π cm^3

(C) 36π cm^3

(D) 108π cm^3

24. Two supplementary angles are in the ratio $2 : 7$. Find both angles.

Your Answer:

25. What is the measure of each interior angle of a regular pentagon?

(A) $72°$

(B) $90°$

(C) $108°$

(D) $120°$

26. A scatter plot shows data points that form a curved (U-shaped) pattern. This is best described as:

(A) positive linear association

(B) negative linear association

(C) nonlinear association

(D) no association

27. A trend line has equation $y = 1.5x + 2$. Using this, what is y when $x = 4$?

(A) 6

(B) 7

(C) 8

(D) 10

28. Data was collected for x-values from 1 to 10. Using the model to predict y when $x = 50$ is an example of:

(A) interpolation

(B) extrapolation

(C) correlation

(D) association

29. 80 students were surveyed about playing sports and playing a musical instrument.

	Plays sport	No sport	Total
Plays instrument	15	25	40
No instrument	30	10	40
Total	45	35	80

Is there an association between playing a sport and playing an instrument? Justify using conditional relative frequencies.

Your Answer

30. A jar has 6 green and 4 yellow balls. One ball is drawn and replaced. Then another is drawn. $P(\text{both green}) = ?$

(A) $\frac{36}{100}$

(B) $\frac{30}{90}$

(C) $\frac{6}{10}$

(D) $\frac{12}{20}$

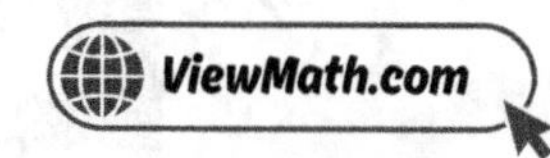

⭐ *End of Practice Test 4* ⭐

Great job finishing the test!

 My Score

I got ___________ out of 30 questions right.

Check your answers in the **Answer Key** at the back of the book.

💡 *Review any questions you missed. That's how we learn!*

📊 **Check Your Score Online!**

Visit **ViewMath Academy** to enter your answers and see which topics you need to review. You can also explore lessons, take quizzes, track your scores, and save your progress!

viewmath.com/score/8.1.FL.19

Or go to *viewmath.com/score* and enter code: 8.1.FL.19

5

Practice Test 5

 30 Questions

✏️ Before You Start ✏️

- ✔ **Read each question carefully** before choosing your answer.
- ✔ **Show your work** on scratch paper when you need to.
- ✔ **Skip hard questions** and come back to them later.
- ✔ **Check your answers** when you're done.
- ✔ **Take your time** — there's no rush!

 ⭐ You've Got This! ⭐

Do your best and show what you know!

1. A calculator shows $\sqrt{144} = 12$. Is $\sqrt{144}$ rational or irrational? Explain.

Your Answer:

2. Between which two consecutive integers does $\sqrt{50}$ lie?

(A) 6 and 7

(B) 7 and 8

(C) 8 and 9

(D) 24 and 26

3. You deposit $200 in a savings account earning 5% compound interest per year. How much more interest do you earn with compound interest compared to simple interest after 3 years?

(A) $0 (they are the same)

(B) $1.53

(C) $5.00

(D) $10.00

4. Study the number line below. Which expression corresponds to the point marked with a star (★)?

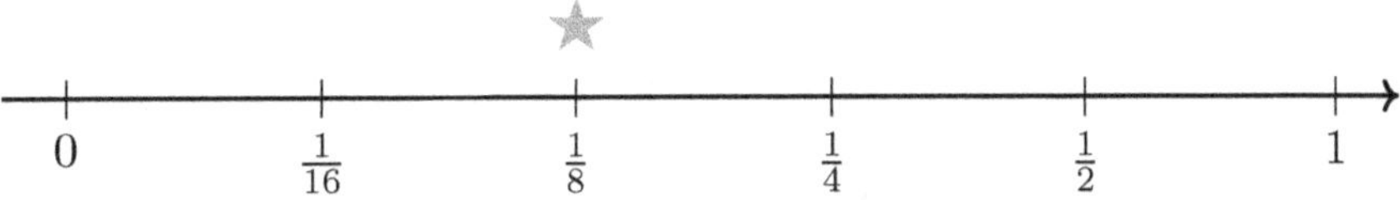

(A) 2^{-2}

(B) 2^{-3}

(C) 2^{-4}

(D) 2^{-1}

5. Which of the following is a perfect square?

(A) 50

(B) 72

(C) 81

(D) 90

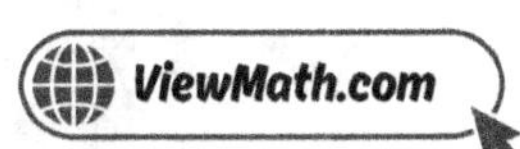

6. Compute $(8 \times 10^{-4})(5 \times 10^{-3})$. Write your answer in scientific notation.

Your Answer:

7. Store A sells apples for \$2 per pound. Store B's prices are shown in the table.

Pounds	3	6	9
Cost (\$)	7.50	15.00	22.50

Which store has the lower unit price?

(A) Store A

(B) Store B

(C) They charge the same price.

(D) Not enough information.

8. A bathtub drains from 40 gallons to 10 gallons in 15 minutes. What is the rate of change?

(A) -2 gal/min

(B) 2 gal/min

(C) -30 gal/min

(D) 30 gal/min

9. Solve $8 - 3x = 2$.

(A) $x = -2$

(B) $x = 2$

(C) $x = 3$

(D) $x = \frac{10}{3}$

10. A rectangle has a perimeter of 48 m. Its length is twice its width. Find the dimensions.

Your Answer:

11. If $f(x) = 2x + 9$, what is $f(5)$?

> Your Answer

12. Function P is given by the table:

x	0	1	2	3
y	4	7	10	13

Function Q: $y = 2x + 5$. Which function has a greater rate of change?

(A) Function P

(B) Function Q

(C) They have the same rate of change.

(D) Cannot be determined.

13. A table shows $(1, 1)$, $(2, 8)$, $(3, 27)$, $(4, 64)$. Is this linear or nonlinear? What pattern do you see?

> Your Answer

14. A gym charges a one-time fee of $60 plus $25 per month. What is the total cost after 8 months?

(A) $200

(B) $260

(C) $285

(D) $480

15. A ball is thrown straight up. Its height over time:

(A) Increases, then decreases — the graph is non-linear.

(B) Increases, then decreases — the graph is linear.

(C) Decreases only — the graph is linear.

(D) Increases only — the graph is nonlinear.

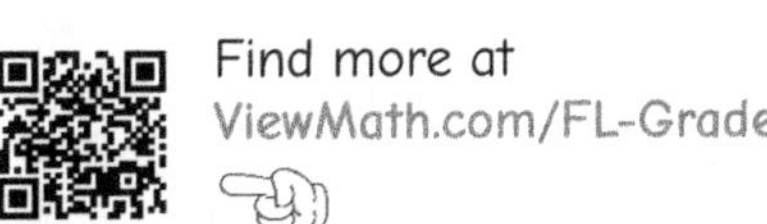

Find more at
ViewMath.com/FL-Grade8

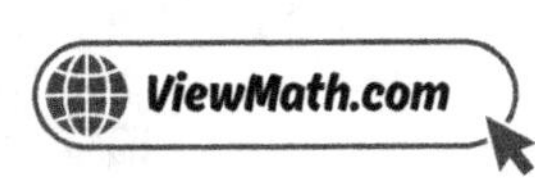

16. Point $N(0, 4)$ is rotated $180°$ around the origin. What are the coordinates of N'?

Your Answer

17. $\triangle ABC \cong \triangle DEF$. If $\angle A = 55°$ and $\angle B = 80°$, what is $\angle F$?

(A) $55°$

(B) $80°$

(C) $45°$

(D) $135°$

18. What is the image of $(-4, 2)$ after a reflection over the y-axis?

(A) $(-4, -2)$

(B) $(4, -2)$

(C) $(4, 2)$

(D) $(2, -4)$

19. Are all squares similar to each other?

(A) No, because they can have different side lengths.

(B) No, because they can have different angles.

(C) Yes, because all squares have equal angles and proportional sides.

(D) Yes, but only if they have the same perimeter.

20. Corresponding angles formed by a transversal and two parallel lines are located:

(A) On opposite sides of the transversal, between the parallel lines

(B) On the same side of the transversal, in matching positions

(C) On opposite sides of the transversal, outside the parallel lines

(D) Adjacent to each other at one intersection

Find more at
ViewMath.com/FL-Grade8

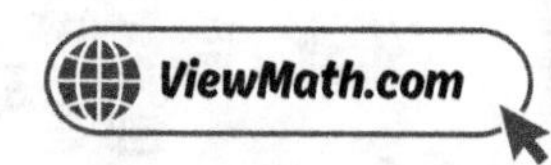

21. *A right triangle has legs 10 and 24. What is the hypotenuse?*

> Your Answer:

22. *What is the distance between $(-2, 1)$ and $(4, 9)$?*

    (A) 10   (B) $\sqrt{10}$

    (C) 14   (D) 8

23. *A traffic cone has radius 8 in and height 24 in. What is its volume? Use $\pi \approx 3.14$*

    (A) $4{,}823.0\ in^3$ (B) $1{,}607.7\ in^3$

    (C) $803.8\ in^3$ (D) $602.9\ in^3$

24. *Two angles are complementary. One angle is 37°. What is the other?*

    (A) 143°   (B) 53°

    (C) 37°   (D) 63°

25. *Each interior angle of a regular polygon is 140°. How many sides does it have?*

    (A) 7   (B) 8

    (C) 9   (D) 10

26. *Which variable typically goes on the x-axis in a scatter plot?*

    (A) The dependent variable   (B) The response variable

    (C) The independent (explanatory) variable   (D) The variable with the largest values

27. A trend line passes through $(2, 10)$ and $(8, 4)$. Find the slope and y-intercept.

Your Answer:

28. Using the model $y = 4x + 7$, for what value of x is $y = 35$?

(A) 7

(B) 6

(C) 8

(D) 10

29. Complete the two-way table and find the relative frequencies for each cell (as fractions of the total).

	Homework Yes	Homework No	Total
Grade A	24	6	
Not Grade A	16	14	
Total			

Your Answer:

30. $P(A) = 0.6$, $P(B) = 0.4$. If A and B are independent, $P(A \text{ and } B) = ?$

(A) 1.0

(B) 0.24

(C) 0.20

(D) 0.10

 End of Practice Test 5 

Great job finishing the test!

 My Score

I got _____________ out of 30 questions right.

*Check your answers in the **Answer Key** at the back of the book.*

💡 *Review any questions you missed. That's how we learn!*

📊 Check Your Score Online!

Visit **ViewMath Academy** to enter your answers and see which topics you need to review. You can also explore lessons, take quizzes, track your scores, and save your progress!

viewmath.com/score/8.1.FL.20

Or go to viewmath.com/score and enter code: 8.1.FL.20

6

Practice Test 6

 30 Questions

---

✏ Before You Start ✏

- ✓ **Read each question carefully** before choosing your answer.
- ✓ **Show your work** on scratch paper when you need to.
- ✓ **Skip hard questions** and come back to them later.
- ✓ **Check your answers** when you're done.
- ✓ **Take your time** — there's no rush!

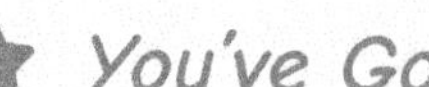 You've Got This! 

Do your best and show what you know!

1. Is $\frac{5}{6}$ rational or irrational? Explain.

Your Answer:

2. A student says $\sqrt{45}$ is between 6 and 7. Is the student correct?

  (A) Yes, because $6^2 = 36$ and $7^2 = 49$.

  (B) No, it is between 5 and 6.

  (C) No, it is between 7 and 8.

  (D) No, $\sqrt{45} = 9$ exactly.

3. Which is a better deal: a \$60 item with a 20% discount, or the same item at a different store for \$50 with no discount?

  (A) The 20% discount, because you pay \$48.

  (B) The \$50 price, because $\$50 < \$60$.

  (C) They cost the same.

  (D) The 20% discount, because you pay \$40.

4. What is the value of $(-2)^0$?

  (A) -2

  (B) 0

  (C) 1

  (D) -1

5. A cube has a volume of 343 cubic inches. What is the length of one edge?

  (A) 7 in.

  (B) 49 in.

  (C) $114.\overline{3}$ in.

  (D) 17 in.

6. What is $7.5 \times 10^8 - 2.5 \times 10^8$?

  (A) 5×10^0

  (B) 5×10^8

  (C) 10×10^8

  (D) 5×10^{16}

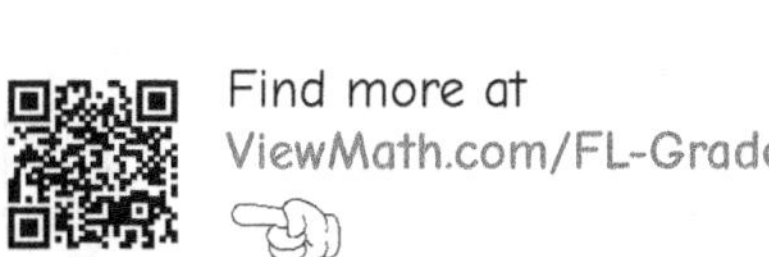

Find more at
ViewMath.com/FL-Grade8

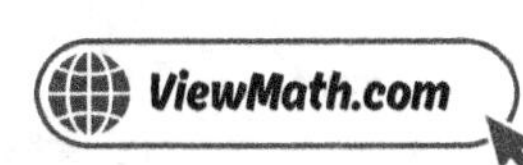

7. A line passes through the origin and the point $(6, 18)$. What is the equation of the line?

   (A) $y = 6x$ (B) $y = 18x$

   (C) $y = 3x$ (D) $y = \frac{1}{3}x$

8. A car's odometer reads 120 miles at 2:00 PM and 270 miles at 5:00 PM. What is the car's speed (rate of change)?

   (A) 45 mph   (B) 50 mph

   (C) 60 mph   (D) 90 mph

9. Solve $2(x + 5) = 3x - 1$.

   (A) $x = 9$ (B) $x = 10$

   (C) $x = 11$ (D) $x = 12$

10. Maria has 15 coins, all nickels and dimes. The coins are worth \$1.10. How many dimes does she have?

   (A) 5   (B) 6

   (C) 7   (D) 8

11. If $f(x) = 7 - x$, what is $f(10)$?

   (A) -3 (B) 3

   (C) 17   (D) -17

12. The graph below shows Function A and Function B. Which function has a greater rate of change?

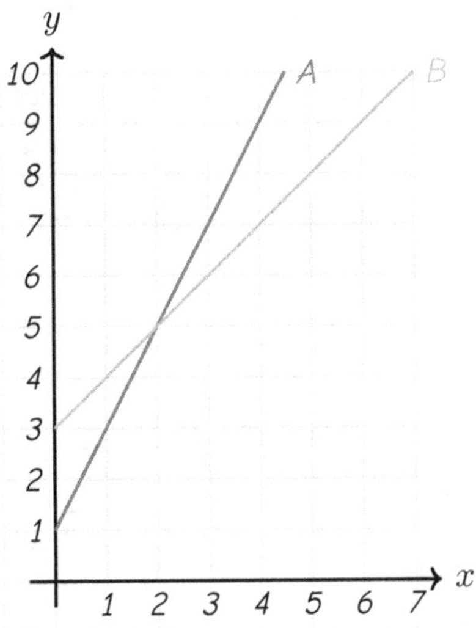

(A) Function A

(B) Function B

(C) They have the same rate of change.

(D) Cannot be determined from the graph.

13. A function increases by 6 for every increase of 1 in x. Is it linear or nonlinear?

Your Answer:

14. A candle is 12 inches tall and burns at 2 inches per hour. What is the function for the candle's height y after x hours?

(A) $y = 2x + 12$

(B) $y = -2x + 12$

(C) $y = 12x - 2$

(D) $y = -12x + 2$

15. A graph shows temperature over a day. From 6 AM to noon, the graph goes up steeply. From noon to 3 PM it goes up slowly. What can you conclude?

(A) The temperature decreased between noon and 3 PM.

(B) The temperature increased faster in the morning than in the afternoon.

(C) The temperature was constant from noon to 3 PM.

(D) The temperature increased at the same rate all day.

16. A student says that a reflection changes the area of a figure. Is this correct?

(A) Yes, because the figure flips.

(B) Yes, because one dimension reverses.

(C) No, because reflections preserve all measurements.

(D) No, but reflections change the perimeter.

17. Two congruent rectangles each have a perimeter of 28 cm. One has a length of 9 cm. What is its width?

Your Answer:

18. A point (a, b) is translated by $(5, -1)$ to get image $(2, 7)$. What is (a, b)?

Your Answer:

19. Which statement about similar figures is always true?

(A) They have equal side lengths.

(B) They have equal angle measures.

(C) They have the same perimeter.

(D) They have the same area.

20. A triangle has angles $90°$, $x°$, and $(x + 10)°$. Find x.

Your Answer:

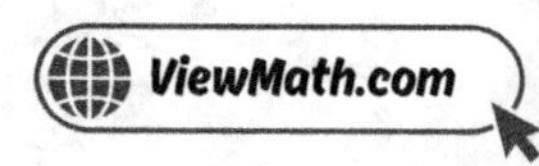

21. A right triangle has legs 1 and 1. What is the exact length of the hypotenuse?

(A) 2

(B) $\sqrt{2}$

(C) 1

(D) $\sqrt{3}$

22. Find the distance between $(0,0)$ and $(2,3)$. Leave your answer in simplest radical form.

Your Answer:

23. A cone has radius 3 m and height 12 m. What is its volume in terms of π?

(A) $108\pi\ m^3$

(B) $36\pi\ m^3$

(C) $12\pi\ m^3$

(D) $324\pi\ m^3$

24. Which pair of angles are complementary?

(A) $45°$ and $135°$

(B) $60°$ and $120°$

(C) $25°$ and $65°$

(D) $90°$ and $90°$

25. How many triangles can a hexagon be divided into by drawing diagonals from one vertex?

(A) 3

(B) 4

(C) 5

(D) 6

26. Data: $(1,2),(2,4),(3,5),(4,8),(5,9)$. The dots appear to go upward from left to right. The association is:

(A) negative

(B) positive

(C) no association

(D) nonlinear

27. *Data:* $(1,3), (2,5), (3,4), (4,6), (5,8)$. *A trend line through* $(1,3)$ *and* $(5,8)$ *has slope:*

(A) $\frac{5}{4}$

(B) $\frac{4}{5}$

(C) 1

(D) 2

28. *What is extrapolation?*

(A) *Making predictions within the range of the data*

(B) *Making predictions far beyond the range of the data*

(C) *Removing outliers from the data*

(D) *Finding the exact equation of the line*

29. *A survey asks students about their favorite sport (soccer or basketball) and grade (7th or 8th). 20 7th-graders chose soccer. This 20 is a:*

(A) *marginal frequency*

(B) *joint frequency*

(C) *relative frequency*

(D) *total frequency*

30. *What is the probability of an impossible event?*

(A) 1

(B) 0.5

(C) 0

(D) -1

 # End of Practice Test 6 

Great job finishing the test!

 My Score

I got ____________ out of 30 questions right.

Check your answers in the **Answer Key** at the back of the book.

💡 Review any questions you missed. That's how we learn!

📊 Check Your Score Online!

Visit **ViewMath Academy** to enter your answers and see which topics you need to review. You can also explore lessons, take quizzes, track your scores, and save your progress!

viewmath.com/score/8.1.FL.21

Or go to viewmath.com/score and enter code: 8.1.FL.21

7

Practice Test 7

☑ *30 Questions*

✏ Before You Start ✏

✔ **Read each question carefully** before choosing your answer.

✔ **Show your work** on scratch paper when you need to.

✔ **Skip hard questions** and come back to them later.

✔ **Check your answers** when you're done.

✔ **Take your time** — there's no rush!

★ You've Got This! ★

Do your best and show what you know!

1. *Give an example of an irrational number between 1 and 2.*

   Your Answer:

2. *Which list is in order from least to greatest?*

   A) $\sqrt{5},\ 2,\ \frac{5}{2}$

   B) $2,\ \sqrt{5},\ \frac{5}{2}$

   C) $\frac{5}{2},\ 2,\ \sqrt{5}$

   D) $2,\ \frac{5}{2},\ \sqrt{5}$

3. *A meal costs \$65. You leave an 18% tip. How much is the tip?*

   Your Answer:

4. *Which expression is equivalent to $\frac{9^3 \cdot 9^2}{9^4}$?*

   A) 9^{-1}

   B) 9^0

   C) 9^1

   D) 9^{24}

5. *A cube-shaped box has a volume of 216 cubic centimeters. What is the edge length of the box?*

   Your Answer:

6. *What is $(3 \times 10^4)(2 \times 10^5)$?*

   A) 6×10^9

   B) 6×10^{20}

   C) 5×10^9

   D) 6×10^1

Find more at
ViewMath.com/FL-Grade8

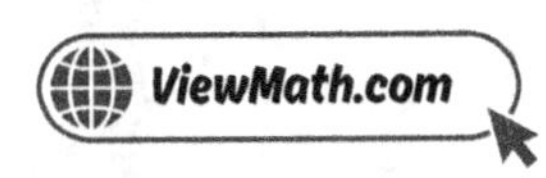

7. A painter paints 3 rooms in 6 hours. At this rate, how many hours will it take to paint 8 rooms?

8. Which pair of points gives a slope of $\frac{1}{2}$?

(A) $(0,0)$ and $(2,4)$

(B) $(1,3)$ and $(5,5)$

(C) $(2,1)$ and $(4,3)$

(D) $(0,1)$ and $(1,4)$

9. Solve $3x + 7 = 22$.

(A) $x = 3$

(B) $x = 5$

(C) $x = 7$

(D) $x = 10$

10. A rectangle's perimeter is 34 cm. The length is 5 cm more than the width. What is the width?

(A) 4 cm

(B) 5 cm

(C) 6 cm

(D) 7 cm

11. If $f(x) = \frac{x+4}{2}$, what is $f(6)$?

Find more at
ViewMath.com/FL-Grade8

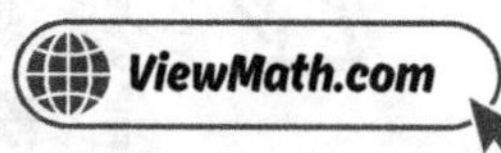
ViewMath.com

12. Function R from the previous table has a rate of change of:

x	0	2	4	6
y	1	9	17	25

(A) 2

(B) 3

(C) 4

(D) 8

13. The equation $A = s^2$ gives the area of a square with side length s. Is this function linear or nonlinear?

(A) Linear, because area is always positive.

(B) Linear, because it has only one variable.

(C) Nonlinear, because the variable is squared.

(D) Nonlinear, because area is measured in square units.

14. A table shows $(0, 8)$, $(3, 20)$, $(6, 32)$. What is the equation?

(A) $y = 3x + 8$

(B) $y = 4x + 8$

(C) $y = 8x + 4$

(D) $y = 6x + 8$

15. A graph of amount of gas in a car over time goes downward. Is the function increasing, decreasing, or constant?

Your Answer:

16. Which transformation flips a figure over a line?

(A) Translation

(B) Rotation

(C) Dilation

(D) Reflection

Find more at
ViewMath.com/FL-Grade8

ViewMath.com

17. Two figures are congruent if one can be mapped onto the other using which of the following?

(A) Only translations

(B) Only dilations

(C) A sequence of rigid transformations

(D) Any single transformation

18. A point is dilated by factor 3 from the origin. If the image is $(9, -15)$, what was the original point?

(A) $(27, -45)$

(B) $(3, -5)$

(C) $(6, -12)$

(D) $(12, -18)$

19. All circles are similar to each other. Why?

(A) They all have the same radius.

(B) One can always be dilated to match the other.

(C) They all have the same area.

(D) Circles cannot be dilated.

20. In a triangle, the angles are $(2x + 5)°$, $(3x)°$, and $(x + 25)°$. Find x.

Your Answer:

21. A right triangle has hypotenuse 17 and one leg 8. What is the other leg?

(A) 9

(B) 15

(C) 25

(D) $\sqrt{225}$

22. What is the distance between $(1, 1)$ and $(4, 5)$?

(A) 25

(B) 7

(C) $\sqrt{5}$

(D) 5

Find more at
ViewMath.com/FL-Grade8

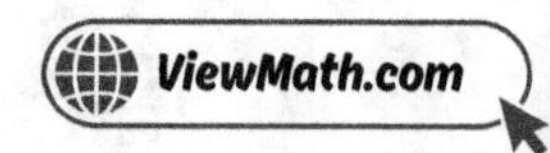

23. *Find the volume of a cylinder with radius 6 cm and height 5 cm. Leave your answer in terms of π.*

Your Answer:

24. *In the figure, two lines cross. What is the measure of angle y?*

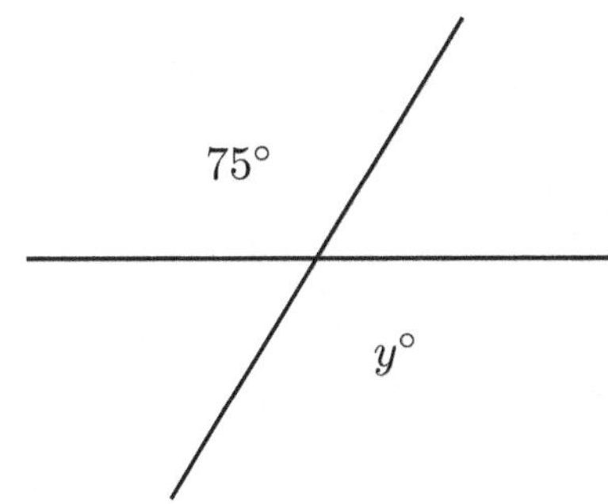

(A) 105°

(B) 15°

(C) 75°

(D) 180°

25. *A regular polygon has each interior angle measuring 160°. How many sides does it have?*

Your Answer:

26. *A scatter plot has a strong upward trend. Which r-value (correlation) is most likely?*

(A) $r = -0.9$

(B) $r = 0.1$

(C) $r = 0.9$

(D) $r = 0$

27. *Data:* $(1, 8), (2, 7), (3, 5), (4, 4), (5, 2)$. *Draw a trend line and write its equation.*

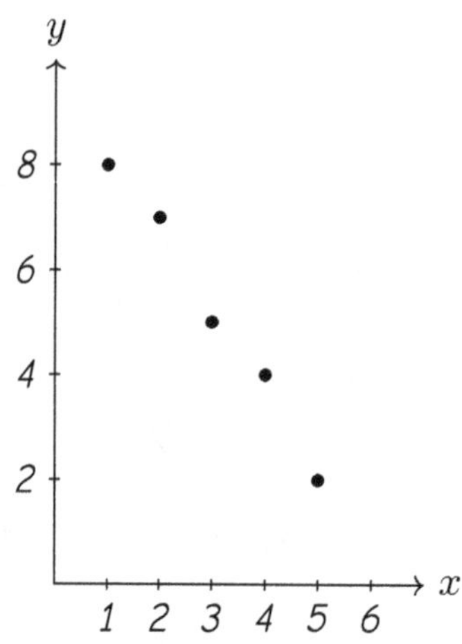

Your Answer:

28. *A linear model is* $y = 2x + 5$. *What does the slope represent?*

(A) The starting value

(B) The rate of change per unit increase in x

(C) The x-intercept

(D) The total value of y

29. *Using the table above, what proportion of pizza lovers are boys?*

(A) $\frac{18}{42}$

(B) $\frac{24}{42}$

(C) $\frac{24}{80}$

(D) $\frac{24}{40}$

30. *A card is drawn from a standard 52-card deck. What is* $P(\text{heart})$?

(A) $\frac{1}{52}$

(B) $\frac{1}{13}$

(C) $\frac{1}{4}$

(D) $\frac{1}{2}$

Find more at
ViewMath.com/FL-Grade8

 # End of Practice Test 7 

Great job finishing the test!

☑ My Score

I got ______________ out of 30 questions right.

Check your answers in the **Answer Key** at the back of the book.

💡 *Review any questions you missed. That's how we learn!*

📊 Check Your Score Online!

Visit **ViewMath Academy** to enter your answers and see which topics you need to review. You can also explore lessons, take quizzes, track your scores, and save your progress!

viewmath.com/score/8.1.FL.22

Or go to *viewmath.com/score* and enter code: 8.1.FL.22

8

Practice Test 8

 30 Questions

✏️ Before You Start ✏️

- ✔ **Read each question carefully** before choosing your answer.
- ✔ **Show your work** on scratch paper when you need to.
- ✔ **Skip hard questions** and come back to them later.
- ✔ **Check your answers** when you're done.
- ✔ **Take your time** — there's no rush!

⭐ You've Got This! ⭐

Do your best and show what you know!

1. Which of the following is a true statement?

   (A) Every square root is irrational.

   (B) Every integer is irrational.

   (C) Every integer is rational.

   (D) Every decimal is irrational.

2. The squares below have the given areas. Estimate the side length of each square to one decimal place.

   A

   B

   C

   Area = 18 sq units   Area = 40 sq units   Area = 72 sq units

   Your Answer:

3. A store buys a backpack for \$24 and marks it up by 75%. What is the selling price?

   Your Answer:

4. Which shows $\frac{1}{2^4}$ written with a negative exponent?

   (A) $(-2)^4$

   (B) 2^{-4}

   (C) -2^4

   (D) 4^{-2}

5. Evaluate $\sqrt[3]{-125}$.

   Your Answer:

6. The bar chart compares the distances of four planets from the Sun. Which planet is approximately 10 times farther from the Sun than Planet A?

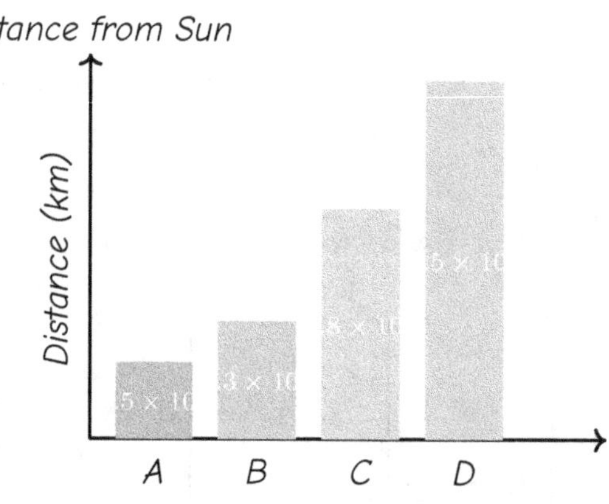

- (A) Planet B
- (B) Planet C
- (C) Planet D
- (D) None of the planets

7. Hose A fills a pool at $y = 15x$ gallons per minute. Hose B fills 100 gallons in 8 minutes. Which fills faster?

- (A) Hose A
- (B) Hose B
- (C) They fill at the same rate.
- (D) Not enough information.

8. A line has a slope of 0. Which of the following best describes the line?

- (A) It is vertical.
- (B) It goes up steeply from left to right.
- (C) It is horizontal.
- (D) It has undefined slope.

9. When solving an equation, the variable disappears and you get $5 = 5$. This means:

- (A) The equation has no solution.
- (B) The equation has exactly one solution, $x = 5$.
- (C) The equation has infinitely many solutions.
- (D) You made a mistake.

10. One number is 4 more than another. Their sum is 36. Find both numbers.

Your Answer

11. The table below shows values for $f(x)$. If $f(x) = 15$, what is x?

x	0	1	2	3	4
$f(x)$	3	6	9	12	15

(A) 2

(B) 3

(C) 4

(D) 5

12. Function R is shown in the table:

x	0	2	4	6
y	1	9	17	25

Function S: $y = 3x + 1$. Which has a greater initial value?

(A) Function R

(B) Function S

(C) They have the same initial value.

(D) Cannot be determined.

13. Give an example of a nonlinear equation.

Your Answer

 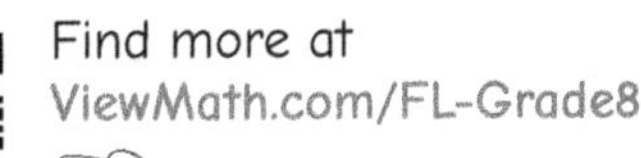
Find more at
ViewMath.com/FL-Grade8

14. The table shows a real-world function. Write the equation and explain what m and b represent.

Hours worked (x)	Pay (y)
0	$50
1	$62
2	$74
3	$86

Your Answer:

15. A student fills a glass with water, drinks half, then refills it. Describe the graph of water level vs. time.

Your Answer:

16. A point (a, b) is reflected over the x-axis, then reflected over the y-axis. What are the final coordinates?

Your Answer:

17. $\triangle PQR \cong \triangle STU$. If $\angle P = 72°$ and $\angle Q = 53°$, what is $\angle U$?

Your Answer:

18. Which transformation maps (x, y) to $(-y, x)$?

(A) Reflection over the x-axis

(B) Reflection over the y-axis

(C) 90° counterclockwise rotation

(D) 180° rotation

19. *Triangle A has sides 6, 8, 10. Triangle B has sides 9, 12, 15. What is the scale factor from A to B?*

Your Answer:

20. *In the figure below, what is the value of y?*

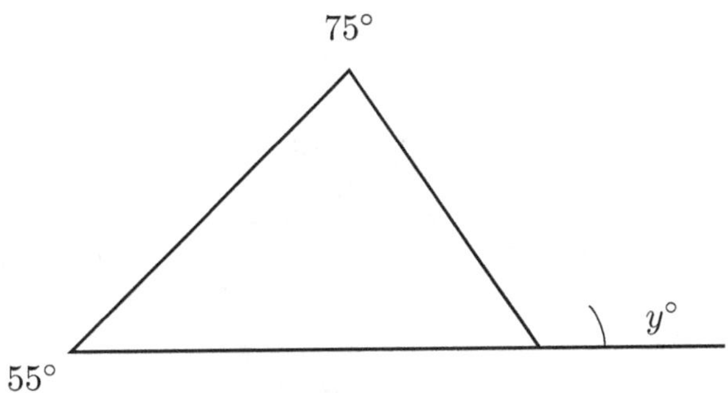

(A) 50°

(B) 105°

(C) 130°

(D) 125°

21. *A rectangular room is 12 ft long and 5 ft wide. What is the diagonal distance across the floor?*

Your Answer:

22. *The distance formula is derived from:*

(A) The area formula for a rectangle

(B) The Pythagorean Theorem

(C) The perimeter formula

(D) The midpoint formula

23. *What is the volume of a sphere with radius 3 cm? Use $\pi \approx 3.14$.*

(A) 28.26 cm^3

(B) 113.04 cm^3

(C) 36π cm^3

(D) 84.78 cm^3

Find more at
ViewMath.com/FL-Grade8

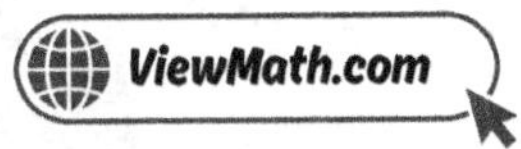

24. Two vertical angles measure $(4x + 8)°$ and $(6x - 12)°$. Find x.

Your Answer:

25. A hexagon has five angles that each measure 125°. What is the sixth angle?

Your Answer:

26. In a scatter plot, several data points bunch together near $(3, 20)$. This grouping is called:

(A) an outlier

(B) a gap

(C) a cluster

(D) a negative association

27. Data: $(1, 2), (2, 4), (3, 7), (4, 8), (5, 11)$. Estimate the slope of a reasonable trend line.

Your Answer:

28. In the model $y = 6x + 12$, if x increases by 3 units, how much does y increase?

(A) 3

(B) 6

(C) 18

(D) 30

29. Using the table from q19, what percentage of boys prefer reading? What percentage of girls prefer reading? Is there an association?

Your Answer:

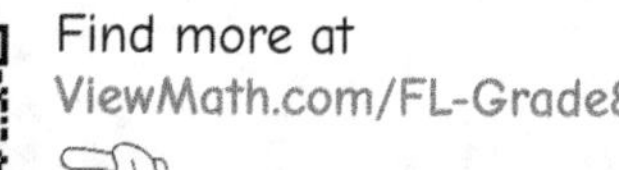

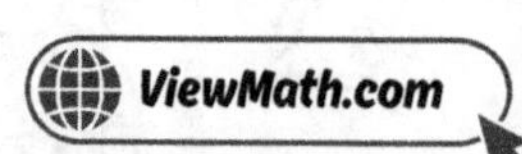

30. $P(event) = \frac{3}{4}$. What is $P(not\ event)$?

(A) $\frac{3}{4}$

(B) $\frac{1}{4}$

(C) $\frac{1}{2}$

(D) 0

 # End of Practice Test 8 

Great job finishing the test!

 My Score

I got _____________ out of 30 questions right.

*Check your answers in the **Answer Key** at the back of the book.*

💡 *Review any questions you missed. That's how we learn!*

📊 Check Your Score Online!

Visit **ViewMath Academy** to enter your answers and see which topics you need to review. You can also explore lessons, take quizzes, track your scores, and save your progress!

viewmath.com/score/8.1.FL.23

Or go to viewmath.com/score and enter code: 8.1.FL.23

Practice Test 9

 30 Questions

 Before You Start 

- ✓ **Read each question carefully** before choosing your answer.
- ✓ **Show your work** on scratch paper when you need to.
- ✓ **Skip hard questions** and come back to them later.
- ✓ **Check your answers** when you're done.
- ✓ **Take your time** — there's no rush!

 You've Got This! 

Do your best and show what you know!

1. Which of the following could NOT be the decimal expansion of a rational number?

   (A) 2.500

   (B) $0.\overline{9}$

   (C) $1.41421356\ldots$ *(non-repeating)*

   (D) $0.\overline{36}$

2. Between which two consecutive integers does $\sqrt{10}$ lie?

   (A) 2 and 3

   (B) 3 and 4

   (C) 4 and 5

   (D) 5 and 6

3. A \$120 pair of shoes is on sale for 25% off. What is the sale price?

   (A) \$30

   (B) \$95

   (C) \$90

   (D) \$25

4. Which expression equals $\frac{1}{9}$?

   (A) 3^{-2}

   (B) 3^{-3}

   (C) 9^{-2}

   (D) $(-3)^2$

5. Between which two consecutive whole numbers does $\sqrt{50}$ lie?

   (A) 6 and 7

   (B) 7 and 8

   (C) 24 and 26

   (D) 8 and 9

6. A factory makes 2.4×10^3 gadgets per hour. How many gadgets does it make in 5×10^2 hours?

   (A) 1.2×10^5

   (B) 1.2×10^6

   (C) 12×10^5

   (D) 7.4×10^3

Find more at
ViewMath.com/FL-Grade8

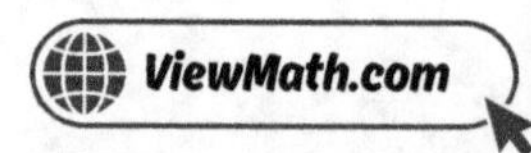

7. A proportional relationship passes through $(8, 12)$. What is y when $x = 20$?

(A) 24

(B) 28

(C) 30

(D) 32

8. The table shows a runner's distance over time.

Time (min)	0	5	10	15	20
Distance (m)	0	40	80	120	160

What is the runner's speed (slope of distance vs. time)?

Your Answer:

9. Solve $4(x - 3) = 2(x + 1)$.

(A) $x = 5$

(B) $x = 7$

(C) $x = 8$

(D) $x = 10$

10. A store sells T-shirts for \$12 and hats for \$8. Use the information in the table to find how many of each were sold.

	Value
Total items sold	15
Total revenue	\$148

Your Answer:

Find more at
ViewMath.com/FL-Grade8

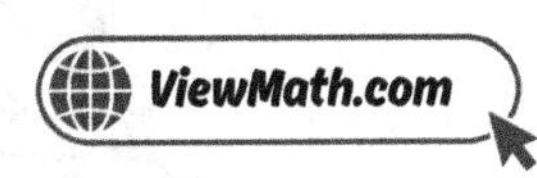

11. If $f(x) = -x + 8$, what is $f(-3)$?

(A) 5

(B) -5

(C) 11

(D) -11

12. Function A: $y = 5x + 3$. Function B: $y = 2x + 10$. Which function has a greater rate of change?

(A) Function A

(B) Function B

(C) They have the same rate of change.

(D) Cannot be determined.

13. Which of these is NOT a linear function?

(A) $y = 0$

(B) $y = 100x$

(C) $y = -x + 5$

(D) $y = \sqrt{x}$

14. A linear function passes through $(0, 5)$ and $(2, 11)$. What is its equation?

(A) $y = 2x + 5$

(B) $y = 3x + 5$

(C) $y = 5x + 3$

(D) $y = 6x + 5$

15. A decreasing graph does NOT necessarily mean:

(A) The output values are getting smaller.

(B) The graph goes downward from left to right.

(C) The output values are negative.

(D) The rate of change is negative.

16. After a translation, a triangle's longest side was originally 8 cm. What is the longest side of the image?

(A) 4 cm

(B) 8 cm

(C) 16 cm

(D) It depends on the direction of the translation.

17. *A pentagon is translated 4 units left and then rotated 90°. Is the image congruent to the original?*

(A) *Yes*

(B) *No, because two transformations were used*

(C) *No, because it was rotated*

(D) *Only if it is a regular pentagon*

18. *Point $(-3, 5)$ is rotated 90° counterclockwise around the origin. What is its image?*

(A) $(5, 3)$

(B) $(-5, -3)$

(C) $(3, -5)$

(D) $(-5, 3)$

19. *A dilation with $k = 1$ produces:*

(A) *A figure twice as large*

(B) *A congruent figure*

(C) *A figure half as large*

(D) *No figure at all*

20. *Triangle ABC has $\angle A = 45°$ and $\angle B = 85°$. What type of triangle is it?*

(A) *Acute*

(B) *Right*

(C) *Obtuse*

(D) *Equilateral*

21. *Is the triangle with sides 9, 40, 41 a right triangle?*

(A) *No, because $9 + 40 \neq 41$.*

(B) *No, because $9^2 + 40^2 \neq 41^2$.*

(C) *Yes, because $9^2 + 40^2 = 41^2$.*

(D) *Yes, because $41 - 40 = 1$.*

22. *What is the distance between $(0, 0)$ and $(3, 4)$?*

(A) 7

(B) 5

(C) $\sqrt{7}$

(D) 12

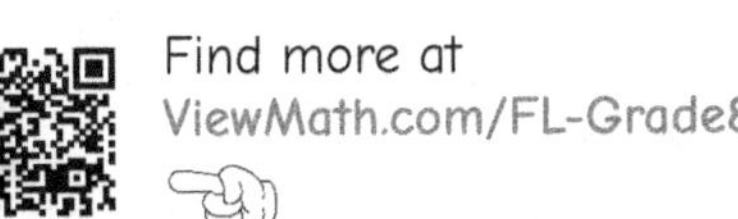

Find more at
ViewMath.com/FL-Grade8

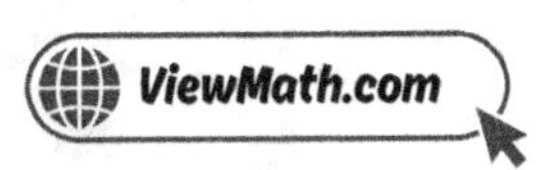

23. A cone has volume 80π cm^3 and radius 4 cm. What is the height?

(A) 5 cm

(B) 15 cm

(C) 10 cm

(D) 20 cm

24. Two angles are supplementary. One angle is 125°. What is the other?

(A) 55°

(B) 35°

(C) 125°

(D) 235°

25. What is the value of x?

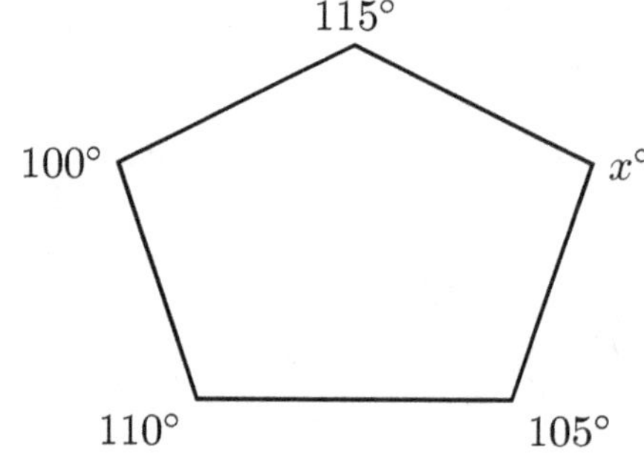

(A) 100°

(B) 110°

(C) 120°

(D) 130°

26. True or false: A scatter plot can show both clustering and outliers at the same time.

(A) True — they describe different features of the data.

(B) False — data has either clusters or outliers, not both.

(C) True — but only if there is no association.

(D) False — outliers are always in clusters.

27. A trend line is drawn through the scatter plot. What is the approximate slope of the line?

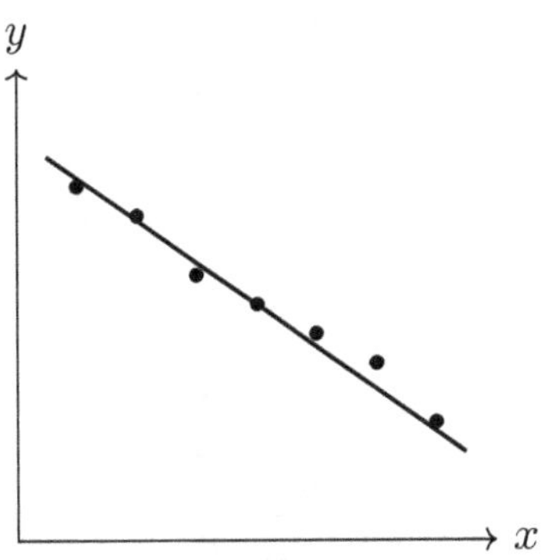

(A) $\approx -\frac{5}{7}$

(B) $\approx \frac{5}{7}$

(C) ≈ -2

(D) ≈ 1

28. A model gives $y = 5x + 20$ where x is weeks and y is savings (dollars). If $x = 8$, what is the predicted savings?

(A) $45

(B) $60

(C) $68

(D) $40

29. Which of these is NOT a question you can answer with a two-way table?

(A) What fraction of boys prefer soccer?

(B) Is there an association between grade and preferred lunch?

(C) What is the mean score on the math test?

(D) How many 8th graders chose art?

30. Explain the difference between independent and dependent events with an example.

Your Answer:

 # End of Practice Test 9 

Great job finishing the test!

☑ My Score

I got ___________ out of 30 questions right.

*Check your answers in the **Answer Key** at the back of the book.*

💡 *Review any questions you missed. That's how we learn!*

📊 Check Your Score Online!

Visit **ViewMath Academy** to enter your answers and see which topics you need to review. You can also explore lessons, take quizzes, track your scores, and save your progress!

viewmath.com/score/8.1.FL.24

Or go to viewmath.com/score and enter code: 8.1.FL.24

Practice Test 10

30 Questions

✏️ Before You Start ✏️

- ✔ **Read each question carefully** before choosing your answer.
- ✔ **Show your work** on scratch paper when you need to.
- ✔ **Skip hard questions** and come back to them later.
- ✔ **Check your answers** when you're done.
- ✔ **Take your time** — there's no rush!

⭐ You've Got This! ⭐

Do your best and show what you know!

1. Look at the Venn diagram below. In which region does $\sqrt{5}$ belong?

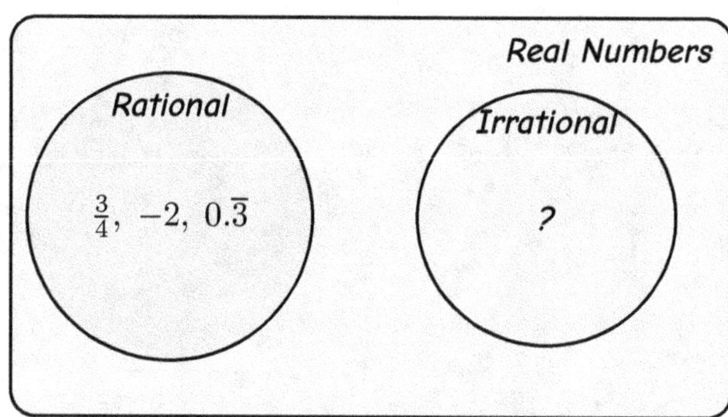

A) Rational region, because $\sqrt{5} = 2.5$

B) Irrational region, because 5 is not a perfect square

C) Rational region, because $\sqrt{5}$ is a square root

D) Neither region, because $\sqrt{5}$ is not a real number

2. Look at the number line below. Which point best represents $\sqrt{6}$?

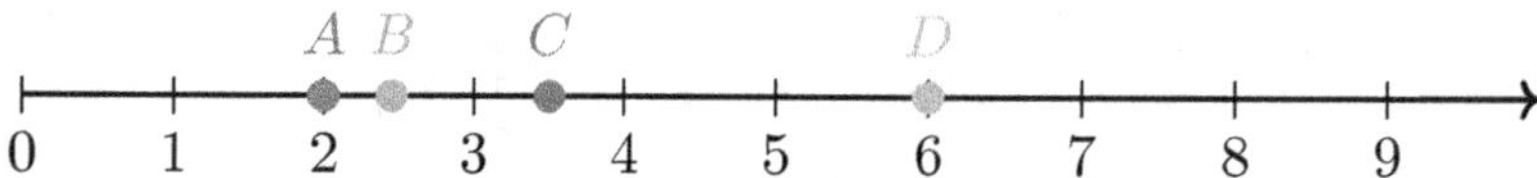

A) Point A

B) Point B

C) Point C

D) Point D

3. A jacket costs $80. Sales tax is 7%. What is the total cost?

A) $85.60

B) $87.00

C) $5.60

D) $74.40

4. Simplify $\left(\frac{3}{4}\right)^{-2}$.

(A) $\frac{9}{16}$

(B) $\frac{-9}{16}$

(C) $\frac{16}{9}$

(D) $\frac{-6}{8}$

5. Which statement about $\sqrt{2}$ is true?

(A) $\sqrt{2}$ is a rational number

(B) $\sqrt{2}$ is an integer

(C) $\sqrt{2}$ is an irrational number

(D) $\sqrt{2} = 1.5$

6. The diagram below shows the input and output of a machine that divides. Find the output value. Write your answer in scientific notation.

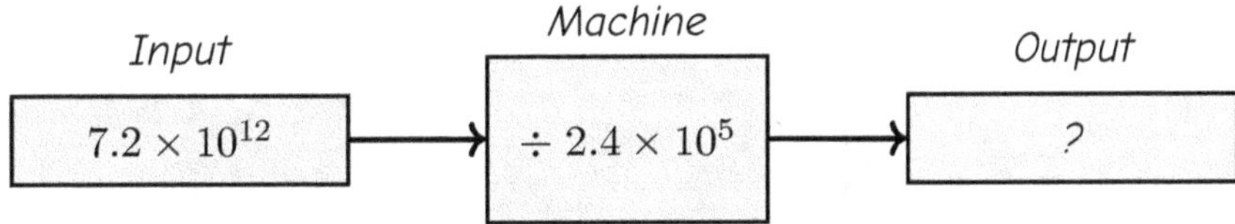

Your Answer:

7. A recipe uses 2 cups of flour for every 5 cookies. How many cups are needed for 30 cookies?

Your Answer:

8. Find the slope through $(3, -1)$ and $(7, 11)$.

Your Answer:

9. How many solutions does $6x + 4 = 6x + 4$ have?

(A) No solution

(B) One solution

(C) Two solutions

(D) Infinitely many solutions

10. Pens cost \$2 and notebooks cost \$5. You buy 8 items for \$25. How many notebooks did you buy?

(A) 2

(B) 3

(C) 4

(D) 5

11. Use the graph of f below to find $f(1) + f(4)$.

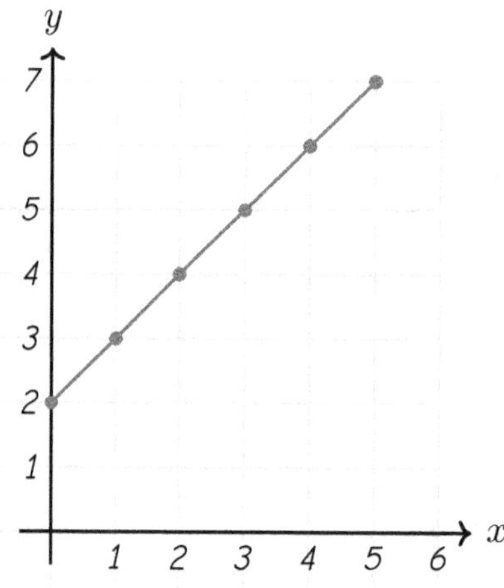

Your Answer

12. Function A: $y = 2x + 100$. Function B: $y = 8x + 10$. At what value of x do the functions have the same output?

(A) $x = 10$

(B) $x = 15$

(C) $x = 18$

(D) $x = 20$

13. The table below shows a function. Calculate the first differences in y and determine whether the function is linear or nonlinear.

x	0	1	2	3	4
y	-3	1	5	9	13

Your Answer:

14. Sam has $120 and spends $15 per day. Which function models the money remaining after x days?

(A) $y = 15x + 120$

(B) $y = -15x + 120$

(C) $y = 120x - 15$

(D) $y = -120x + 15$

15. Two distance-time graphs start at the same point. Graph A is steeper than Graph B. What can you say?

(A) Object A is farther from the start.

(B) Object A is traveling faster than Object B.

(C) Object B is traveling faster than Object A.

(D) Both objects travel at the same speed.

16. Point $M(5, -3)$ is reflected over the y-axis. What are the coordinates of M'?

Your Answer:

17. A student says two circles with radius 5 cm are always congruent. Is this correct?

(A) No, circles cannot be congruent.

(B) No, they need the same center too.

(C) Yes, because they have the same radius.

(D) Yes, but only if they are in the same position.

18. *Point M is shown on the coordinate plane at $(-2, 5)$. Find the coordinates of M' after a 90° counterclockwise rotation around the origin.*

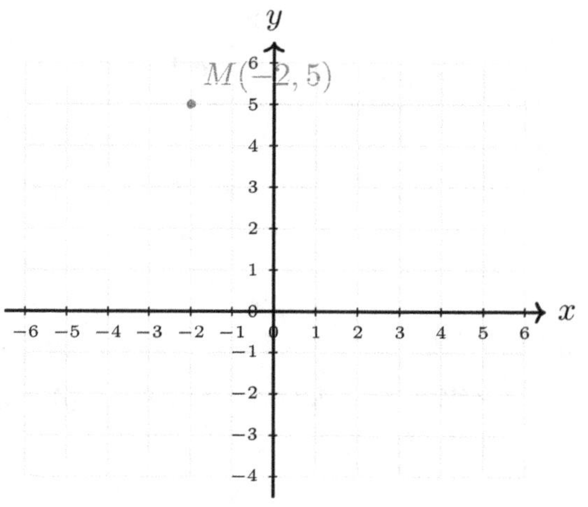

Your Answer:

19. *Two similar triangles have a scale factor of $\frac{1}{3}$ from the larger to the smaller. If the larger triangle's perimeter is 36 cm, what is the smaller triangle's perimeter?*

(A) 12 cm

(B) 108 cm

(C) 9 cm

(D) 18 cm

20. *Two parallel lines are cut by a transversal. If one co-interior (same-side interior) angle is 135°, what is the other?*

(A) 135°

(B) 45°

(C) 55°

(D) 225°

21. In a right triangle, the legs are 6 and 8. What is the hypotenuse?

(A) 14

(B) 10

(C) 48

(D) $\sqrt{48}$

22. Two points are at $(3, 7)$ and $(3, -5)$. What is the distance between them?

(A) 2

(B) 12

(C) -12

(D) $\sqrt{12}$

23. A cylindrical can has radius 3 in and height 8 in. How much soup can it hold? Use $\pi \approx 3.14$.

Your Answer

24. Two angles form a right angle as shown. What is x?

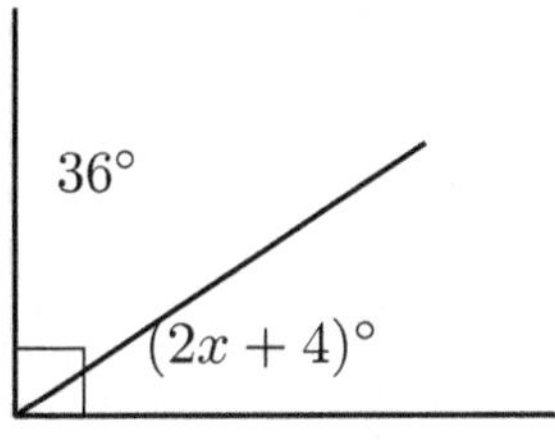

(A) 43

(B) 25

(C) 27

(D) 30

25. Which polygon has an angle sum of 900°?

(A) Pentagon

(B) Hexagon

(C) Heptagon

(D) Octagon

Find more at
ViewMath.com/FL-Grade8

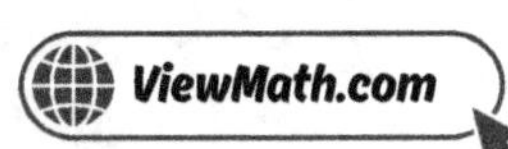

26. A scatter plot of age (x) versus height (y) for people aged 0–80 would likely show what shape?

Your Answer:

27. When drawing an informal line of best fit, about how many data points should be above the line?

(A) All of them

(B) None of them

(C) About half

(D) Exactly one

28. A scatter plot with trend line $y = 2x + 1$ is shown. What is the predicted value at $x = 4$?

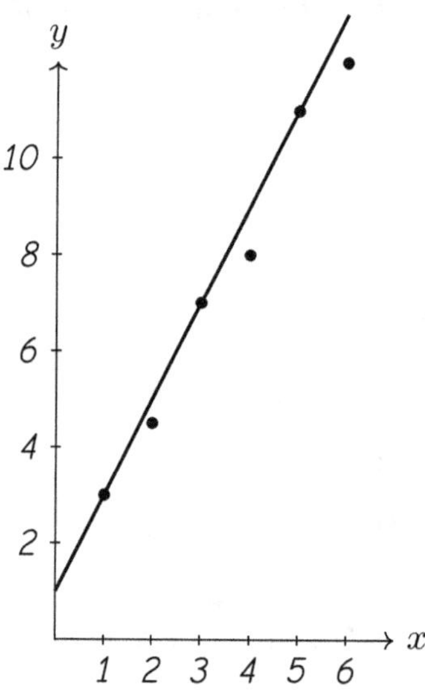

(A) 7

(B) 8

(C) 9

(D) 10

29.

	Pizza	Tacos	Total
Boys	24	16	40
Girls	18	22	40
Total	42	38	80

What is the relative frequency of boys who prefer pizza out of all 80 students?

(A) 0.30

(B) 0.60

(C) 0.20

(D) 0.525

30. A bag has 5 red and 3 blue marbles. One marble is drawn and replaced. Then another is drawn. What is $P(red, then blue)$?

(A) $\frac{15}{64}$

(B) $\frac{15}{56}$

(C) $\frac{5}{8}$

(D) $\frac{3}{8}$

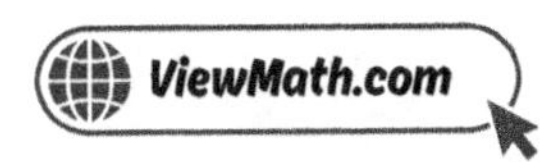

 # End of Practice Test 10 

Great job finishing the test!

☑ My Score

I got _____________ out of 30 questions right.

*Check your answers in the **Answer Key** at the back of the book.*

💡 *Review any questions you missed. That's how we learn!*

📊 Check Your Score Online!

Visit **ViewMath Academy** to enter your answers and see which topics you need to review. You can also explore lessons, take quizzes, track your scores, and save your progress!

viewmath.com/score/8.1.FL.25

Or go to viewmath.com/score and enter code: 8.1.FL.25

Answer Key & Explanations

Answer Key

First try each test on your own, then check your work here.

✓ Practice Test 1 — Answer Key

1 False   **2** A   **3** B   **4** D   **5** B   **6** 317 times   **7** A   **8** B   **9** $x =$

10 80   **11** C

12 Function A: rate $= 3$, initial $= 20$. Function B: rate $= 5$, initial $= 5$. At $x = 10$, $B = 55$ is greater than $A = 50$.

13 Yes   **14** B   **15** Constant   **16** C   **17** Reflection over the y-axis   **18** $(-5, -2)$

19 B   **20** B   **21** B   **22** C   **23** A   **24** B   **25** B   **26** B   **27** B

28 At $x = 5$: $y = 30$. At $x = 30$: $y = -45$. The prediction at $x = 5$ is more reliable because it is within the data ran

29 C   **30** $\frac{4}{52} \times \frac{3}{51} = \frac{12}{2652} = \frac{1}{221}$

💡 Time to Learn! 💡

Review the explanations below, **especially for the questions you missed**.

Understanding why each answer is correct builds stronger problem-solving skills.

Tip: Circle any questions you got wrong, then read their explanation carefully.

📖 Practice Test 1 — Detailed Explanations

1 $\frac{22}{7} \approx 3.142857\ldots$ is a rational approximation of π, but $\pi = 3.14159265\ldots$ is irrational. They are close but not equal.

2 The point is between 5 and 6, closer to 5. $\sqrt{28} \approx 5.29$, which matches this position. $\sqrt{33} \approx 5.74$ would be closer to 6, and $\sqrt{40} \approx 6.32$ and $\sqrt{55} \approx 7.42$ would be beyond 6.

3 Tax $= 65 \times 0.08 = 5.20$. Total $= 65 + 5.20 = \$70.20$.

4 $7^5 \cdot 7^{-5} = 7^{5+(-5)} = 7^0 = 1$.

5 If $s = 8$, then $A = 8^2 = 64$. If $e = 4$, then $V = 4^3 = 64$. Since $A = V = 64$, the pair $s = 8, e = 4$ works.

6 $\frac{1.9 \times 10^{27}}{6 \times 10^{24}} = \frac{1.9}{6} \times 10^3 \approx 0.317 \times 10^3 = 317$.

7 Table: $k = \frac{6}{1} = 6$. Equation: $k = 5$. The table has the larger constant of proportionality.

8 The slope equals the cost per text, which is $\$0.15$.

9 Add $5x$: $14 = 4x + 2$. Subtract 2: $12 = 4x$. Divide by 4: $x = 3$.

10 $a + s = 200$ and $8a + 5s = 1240$. From first: $s = 200 - a$. Substitute: $8a + 5(200 - a) = 1240$, so $3a + 1000 = 1240$, $3a = 240$, $a = 80$.

11 $g(x) = 0$ where the line crosses the x-axis. The graph crosses at $x = 4$.

12 A: slope 3, $b = 20 \Rightarrow A(10) = 3(10) + 20 = 50$. B: slope 5, $b = 5 \Rightarrow B(10) = 5(10) + 5 = 55$. Function B is greater at $x = 10$.

Find more at
ViewMath.com/FL-Grade8

13 $y = -12$ is a horizontal line, which can be written as $y = 0x + (-12)$ ($m = 0$, $b = -12$). It is linear.

14 The rate is \$3 per mile ($m = 3$) and the flat fee is \$4 ($b = 4$). So $y = 3x + 4$.

15 A straight line has a constant slope, meaning the rate of change is the same everywhere along the line.

16 A translation moves the figure to a new position. Side lengths, angle measures, and parallelism are all preserved.

17 Each x-coordinate is negated while y stays the same: $(1, 1) \rightarrow (-1, 1)$, etc. This is a reflection over the y-axis.

18 $180°$ rotation: $(x, y) \rightarrow (-x, -y)$. So $(5, 2) \rightarrow (-5, -2)$.

19 Divide the image coordinates by the original: $\frac{12}{4} = 3$ and $\frac{-6}{-2} = 3$. The scale factor is 3.

20 Alternate interior angles are equal: $4x = 80$, so $x = 20$.

21 $c^2 = 5^2 + 5^2 = 25 + 25 = 50$, so $c = \sqrt{50}$.

22 $d = \sqrt{(6 - 2)^2 + (-2 - (-5))^2} = \sqrt{16 + 9} = \sqrt{25} = 5$.

23 A cone is exactly $\frac{1}{3}$ the volume of a cylinder with the same base and height: $\frac{450\pi}{3} = 150\pi \text{ cm}^3$.

24 If the complement is $17°$, the original angle is $90 - 17 = 73°$.

25 Sum $= 360°$. Fourth angle $= 360 - 85 - 110 - 75 = 90°$.

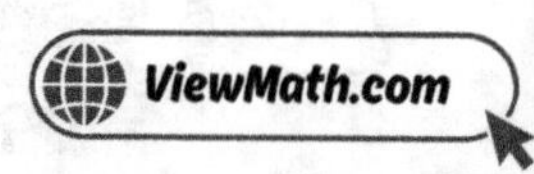

26 The dots go from upper-left to lower-right in a roughly straight pattern. This is a negative linear association.

27 Outliers should generally be ignored when drawing the line of best fit, as they don't represent the overall trend.

28 $y = -3(5) + 45 = 30$ (interpolation). $y = -3(30) + 45 = -45$ (extrapolation — unreliable and gives a negative value that may not make sense).

29 $\frac{25}{40} = 0.625 = 62.5\%$.

30 First ace: $\frac{4}{52}$. Second ace (without replacement): $\frac{3}{51}$. Multiply: $\frac{12}{2652} = \frac{1}{221}$.

✅ Practice Test 2 — Answer Key

1 C

2 $\sqrt{5} \approx 2.2$, placed at the 2nd tick after 2.0

3 $76.32

4 D

5 B

6 D

7 C

8 0.1 miles per minute

9 C

10 5

11 B

12 7

13 Nonlinear

14 4

15 C

16 C

17 B

18 A

19 108 cm^2

20 B

21 B

22 9

23 C

24 False

25 B

26 Positive linear association. Outlier at approximately $(1, 7)$ — far above the trend.

27 C

28 C

29 B

30 C

Find more at
ViewMath.com/FL-Grade8

💡 Time to Learn! 💡

*Review the explanations below, **especially for the questions you missed**.*

Understanding why each answer is correct builds stronger problem-solving skills.

Tip: *Circle any questions you got wrong, then read their explanation carefully.*

📖 Practice Test 2 — Detailed Explanations

1 $P = \sqrt{2}$ is irrational (since 2 is not a perfect square) and $Q = 3$ is an integer, which is rational.

2 $2.2^2 = 4.84$ and $2.3^2 = 5.29$. Since 5 is closer to 4.84, $\sqrt{5} \approx 2.24$, which is near the 2nd to 3rd tick mark after 2.0.

3   Discount: $90 \times 0.20 = 18$. Sale price: $90 - 18 = 72$. Tax: $72 \times 0.06 = 4.32$. Final price: $72 + 4.32 = \$76.32$.

4   A negative exponent means reciprocal: $4^{-3} = \frac{1}{4^3} = \frac{1}{64}$.

5 $x = \pm\sqrt{\frac{9}{16}} = \pm\frac{3}{4}$, since $\left(\frac{3}{4}\right)^2 = \frac{9}{16}$.

6 $9.1 - 8.6 = 0.5$, so 0.5×10^5. In proper notation: 5×10^4. Both representations are equivalent.

7 $k = \frac{y}{x} = \frac{20}{5} = 4$.

8 $m = \frac{5-2}{50-20} = \frac{3}{30} = 0.1$ miles per minute.

9   Subtract 3: $\frac{x}{4} = 4$. Multiply by 4: $x = 16$.

Find more at
ViewMath.com/FL-Grade8

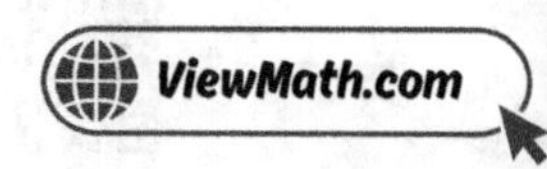

10 $a + b = 12$ and $5a + 10b = 85$. From first: $a = 12 - b$. Substitute: $5(12 - b) + 10b = 85$, so $60 + 5b = 85$, $5b = 25$, $b = 5$.

11 $f(2) = 10 - 3(2) = 10 - 6 = 4$ pieces remaining.

12 The initial value is the output when $x = 0$, which is 7.

13 Check the y-differences: $2 - 1 = 1$, $4 - 2 = 2$, $5 - 4 = 1$, $7 - 5 = 2$. The differences are not all the same $(1, 2, 1, 2)$, so the rate of change is not constant. The function is nonlinear.

14 $\frac{25-9}{6-2} = \frac{16}{4} = 4$.

15 A straight line means linear. Going upward means increasing.

16 Reflecting over the y-axis flips the sign of the x-coordinate while keeping y the same: $(7, -2) \to (-7, -2)$.

17 Equal angles make triangles similar, but not necessarily congruent — their sides could be different lengths.

18 First translate: $(1 + 3, 4 + (-2)) = (4, 2)$. Then reflect over the y-axis: $(4, 2) \to (-4, 2)$.

19 Area scales by k^2: $12 \times 3^2 = 12 \times 9 = 108$ cm^2.

20 Alternate interior angles formed by parallel lines and a transversal are equal: $72°$.

21 $c^2 = 8^2 + 15^2 = 64 + 225 = 289$, so $c = \sqrt{289} = 17$.

22 Same x-coordinates, so distance $= |7 - (-2)| = 9$.

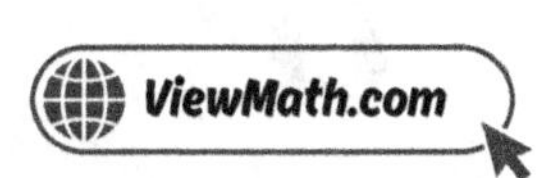

23. $V = \pi r^2 h = \pi(49)(3) = 147\pi \ m^3.$

24. Vertical angles are always equal, not supplementary. (They are supplementary only when each is $90°$.)

25. Sum of hexagon angles $= 720°$. $x = 720 - 130 - 110 - 125 - 130 - 115 = 110°$.

26. Direction: positive (up-right). Shape: linear. The point $(1, 7)$ is an outlier (most students who studied 1 hour scored around 3, not 7).

27. A line of best fit (trend line) is a straight line drawn to approximate the overall trend, coming close to most points.

28. A slope of 0 means no change in y as x changes; the line is horizontal.

29. 60% of boys chose band vs. 45% of girls. The difference in conditional proportions shows an association.

30. 2 coin outcomes $\times$ 6 die outcomes $= 12$ total outcomes.

✅ Practice Test 3 — Answer Key

 C    C    B    B    C    A    A    A    B    C

 B    B   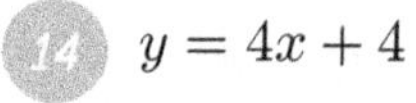 B    $y = 4x + 4$  B    B    A    A    C

 B    B    10    C    B   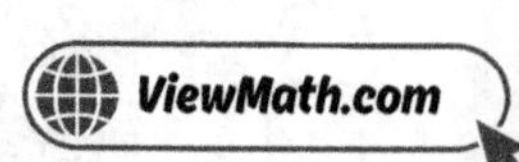 C

26. Yes. Example: population growth over time may curve upward (exponential), showing a positive but nonlinear ass

27. B   28. B   29. B   30. C

Time to Learn!

*Review the explanations below, **especially for the questions you missed.***

Understanding why each answer is correct builds stronger problem-solving skills.

Tip: *Circle any questions you got wrong, then read their explanation carefully.*

📖 Practice Test 3 — Detailed Explanations

1. $0.\overline{81}$ is a repeating decimal, so it can be written as a fraction: $0.\overline{81} = \frac{81}{99} = \frac{9}{11}$. The others are irrational.

2. $5.4^2 = 29.16$ and $5.5^2 = 30.25$. Since 30 is very close to 30.25, $\sqrt{30} \approx 5.5$.

3. $I = 2000 \times 0.06 \times 2 = 240$. Total $= 2000 + 240 = \$2{,}240$.

4. $\frac{8^6}{8^6} = 8^{6-6} = 8^0 = 1$. The expression equals 8^0.

5. Side $= \sqrt{196} = 14$ cm, since $14 \times 14 = 196$.

6. Divide the coefficients: $\frac{8}{2} = 4$. Subtract the exponents: $10^{7-3} = 10^4$. Answer: 4×10^4.

7. Machine A: 12 bottles/hour. Machine B: $\frac{50}{5} = 10$ bottles/hour. Machine A is faster.

8. $m = \frac{1-7}{5-2} = \frac{-6}{3} = -2$.

9. Subtract 1.5: $0.5x = 2.5$. Divide by 0.5: $x = 5$.

  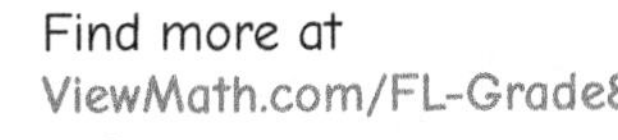
Find more at
ViewMath.com/FL-Grade8
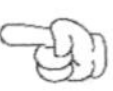

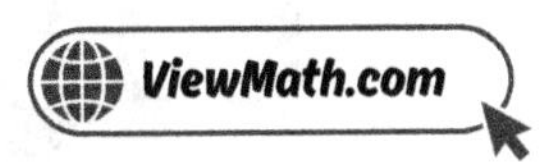

10 $60t + 80t = 420$, so $140t = 420$, $t = 3$ hours.

11 $f(0) = -1$ and $f(2) = 7$. So $f(0) + f(2) = -1 + 7 = 6$.

12 $|-4| = 4 > |-1| = 1$, so Function B decreases faster.

13 $7 - 3 = 4$, $11 - 7 = 4$, $15 - 11 = 4$. The constant change of 4 proves this is linear.

14 Slope: $\frac{24-4}{5-0} = \frac{20}{5} = 4$. y-intercept: 4. Equation: $y = 4x + 4$.

15 A downward straight line means the quantity decreases at a constant rate — water is leaking steadily.

16 The 90° CCW rule is $(x, y) \rightarrow (-y, x)$. So $(-1, 4) \rightarrow (-4, -1)$.

17 Each vertex of Figure A is shifted 6 units right to get Figure B: $(1, 1) \rightarrow (7, 1)$, $(4, 1) \rightarrow (10, 1)$, etc. Since a translation preserves size and shape, the quadrilaterals are congruent.

18 90° CCW: $(x, y) \rightarrow (-y, x)$. So $(0, -5) \rightarrow (5, 0)$.

19 Multiply the side by the scale factor. $3 \times 4 = 12$ cm.

20 $180 - 54 - 54 = 72°$.

21 The diagonal of a square with side s is $s\sqrt{2}$. So the distance is $90\sqrt{2} \approx 127.3$ ft.

22 Horizontal: $4 - (-4) = 8$. Vertical: $4 - (-2) = 6$. $d = \sqrt{8^2 + 6^2} = \sqrt{64 + 36} = \sqrt{100} = 10$.

23 $V = \frac{4}{3}\pi(3r)^3 = \frac{4}{3}\pi(27r^3) = 27 \times \frac{4}{3}\pi r^3$. The volume is 27 times as large.

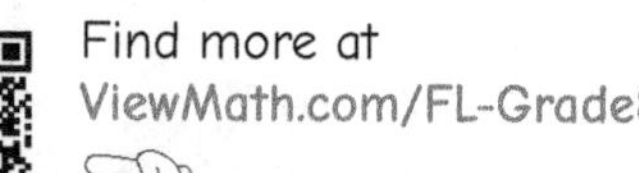
Find more at
ViewMath.com/FL-Grade8

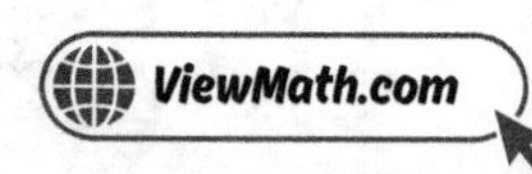

24 $180 - 90 = 90°$. A 90° angle is its own supplement.

25   As n increases, $\frac{(n-2)\times 180}{n}$ approaches 180° but never reaches it.

26   An association can be positive (both increase) and nonlinear (the trend is curved, not straight).

27   A straight line of best fit only works for data with a linear pattern. Curved data needs a different model.

28 $y = -4(10) + 100 = -40 + 100 = 60$.

29   The intersection of Girls row and Dog column is 12.

30 $P(6) = \frac{1}{6}$. Independent: $P = \frac{1}{6} \times \frac{1}{6} = \frac{1}{36}$.

✅ Practice Test 4 — Answer Key

1 D   **2** ≈ 8.2 **3** B   **4** 1.2   **5** B   **6** A   **7** D   **8** 0   **9** C

10 C   **11** C   **12** $A = 21,\ B = 21$ **13** C   **14** $y = 3x$ **15** C   **16** B   **17** C

18 B   **19** A   **20** A   **21** B   **22** C   **23** B   **24** 40° and 140°   **25** C   **26** C

27 C   **28** B

29   Yes. Of instrument players, $\frac{15}{40} = 37.5\%$ play a sport. Of non-instrument players, $\frac{30}{40} = 75\%$ play a sport. The large differe.

30   A

Find more at
ViewMath.com/FL-Grade8

📖 Practice Test 4 — Detailed Explanations

1. $\sqrt{7}$ is irrational because 7 is not a perfect square. The other choices are all rational: $\frac{5}{8} = 0.625$, $0.\overline{6} = \frac{2}{3}$, and $\sqrt{9} = 3$.

2. $8.2^2 = 67.24$ and $8.3^2 = 68.89$. Since 68 is between these and closer to 68.89, $\sqrt{68} \approx 8.2$.

3. $I = Prt$, so $r = \frac{I}{Pt} = \frac{96}{800 \times 4} = \frac{96}{3200} = 0.03 = 3\%$.

4. $5^0 = 1$ and $5^{-1} = \frac{1}{5} = 0.2$. So $1 + 0.2 = 1.2$.

5. $7 \times 7 = 49$, so $\sqrt{49} = 7$.

6. Same exponent, so add the coefficients: $5.2 + 3.8 = 9.0$. Answer: 9×10^6.

7. If the relationship passes through $(1, 5)$ and $(3, 15)$, then $k = 5$. But $5 \times 2 = 10 \neq 13$, so $(2, 13)$ does not fit.

8. $m = \frac{4-4}{8-0} = \frac{0}{8} = 0$. The line is horizontal.

9. Multiply by 3: $2x - 1 = 15$. Add 1: $2x = 16$. Divide by 2: $x = 8$.

Find more at
ViewMath.com/FL-Grade8

10. $x + y = 25$ and $x - y = 7$. Add: $2x = 32$, $x = 16$. The larger number is 16.

11. $h(5) = 2(5)^2 = 2(25) = 50$.

12. $A(3) = 2(3) + 15 = 21$. $B(3) = 5(3) + 6 = 21$. They are equal at $x = 3$.

13. $2 - 1 = 1$, $5 - 2 = 3$, $10 - 5 = 5$, $17 - 10 = 7$. The differences change, so the rate of change is not constant — nonlinear.

14. Slope: $\frac{12-3}{4-1} = \frac{9}{3} = 3$. Using $(1, 3)$: $3 = 3(1) + b \Rightarrow b = 0$. Equation: $y = 3x$.

15. Filling: water level rises (increasing). Soaking: level stays the same (constant). Draining: level drops (decreasing).

16. Each vertex has its x-coordinate negated while y stays the same: $P(1, 1) \to P'(-1, 1)$, $Q(4, 1) \to Q'(-4, 1)$, $R(2, 4) \to R'(-2, 4)$. This is a reflection over the y-axis.

17. Both rectangles have dimensions 6×10. A rotation maps one onto the other, so they are congruent.

18. Add the translation: $(-6 + 4,\ 2 + (-5)) = (-2, -3)$.

19. Congruent figures are a special case of similar figures with $k = 1$. All congruent figures are similar, but not all similar figures are congruent.

20. An exterior angle and its adjacent interior angle are supplementary: $180 - 140 = 40°$.

21. $7^2 + 24^2 = 49 + 576 = 625 = 25^2$. Since $a^2 + b^2 = c^2$, it is a right triangle.

22. $d = \sqrt{(2 - (-1))^2 + (6 - 2)^2} = \sqrt{9 + 16} = \sqrt{25} = 5$.

Find more at
ViewMath.com/FL-Grade8

23 $V = \frac{1}{3}\pi(4^2)(9) = \frac{1}{3}\pi(144) = 48\pi \ cm^3.$

24 $2x + 7x = 180$, so $9x = 180$ and $x = 20$. Angles: $40°$ and $140°$.

25 Sum $= (5 - 2) \times 180 = 540°$. Each angle $= 540/5 = 108°$.

26 A curved pattern is a nonlinear association. The data has a trend, but it is not a straight line.

27 $y = 1.5(4) + 2 = 6 + 2 = 8.$

28 $x = 50$ is far outside the data range (1 to 10), so this is extrapolation.

29 Students who do NOT play an instrument are much more likely to play a sport (75% vs. 37.5%). This difference in conditional percentages signals an association.

30 With replacement: $\frac{6}{10} \times \frac{6}{10} = \frac{36}{100} = \frac{9}{25}.$

✅ Practice Test 5 — Answer Key

1 Rational **2** B **3** B **4** B **5** C **6** 4×10^{-6} **7** A **8** A **9** B

10 Length $= 16$ m, width $= 8$ m **11** 19 **12** A **13** Nonlinear; each output is x^3. **14** B

15 A **16** $(0, -4)$ **17** C **18** C **19** C **20** B

21 26 **22** A **23** B **24** B **25** C **26** C **27** Slope $= -1$; y-intercept $= 12$ **28** A

29 Totals: Grade A $= 30$, Not A $= 30$, HW Yes $= 40$, HW No $= 20$, Grand Total $= 60$. Relative frequencies: $\frac{24}{60} = 0.40,$

Find more at
ViewMath.com/FL-Grade8

B

💡 Time to Learn! 💡

Review the explanations below, **especially for the questions you missed**.

Understanding why each answer is correct builds stronger problem-solving skills.

Tip: Circle any questions you got wrong, then read their explanation carefully.

📖 Practice Test 5 — Detailed Explanations

1. $\sqrt{144} = 12$ exactly, because $12^2 = 144$. Since 12 is an integer, $\sqrt{144}$ is rational.

2. $7^2 = 49$ and $8^2 = 64$. Since $49 < 50 < 64$, we have $7 < \sqrt{50} < 8$.

3. Simple: $I = 200 \times 0.05 \times 3 = \$30$. Compound: Year 1: $200 \times 0.05 = 10 \to \$210$. Year 2: $210 \times 0.05 = 10.50 \to \$220.50$. Year 3: $220.50 \times 0.05 = 11.025 \to \$231.525$. Compound interest $= \$31.53$. Difference: $31.53 - 30 = \$1.53$.

4. The star is at $\frac{1}{8}$. Since $\frac{1}{8} = \frac{1}{2^3} = 2^{-3}$, the answer is 2^{-3}.

5. $81 = 9^2$, so 81 is a perfect square.

6. $8 \times 5 = 40$ and $10^{-4} \times 10^{-3} = 10^{-7}$. Then $40 \times 10^{-7} = 4 \times 10^{-6}$.

7. Store A: $k = 2$ per pound. Store B: $k = \frac{7.50}{3} = 2.50$ per pound. Store A is cheaper.

8. $\frac{10-40}{15} = \frac{-30}{15} = -2$ gal/min. The negative sign shows the water level is decreasing.

Find more at
ViewMath.com/FL-Grade8

9 Subtract 8: $-3x = -6$. Divide by -3: $x = 2$.

10 $2l + 2w = 48$ and $l = 2w$. Substitute: $2(2w) + 2w = 48$, $6w = 48$, $w = 8$. Then $l = 16$.

11 $f(5) = 2(5) + 9 = 10 + 9 = 19$.

12 Function P's rate of change: $\frac{7-4}{1-0} = 3$. Function Q's slope is 2. Since $3 > 2$, Function P grows faster.

13 $1^3 = 1$, $2^3 = 8$, $3^3 = 27$, $4^3 = 64$. The differences $(7, 19, 37)$ are not constant, confirming nonlinear.

14 $y = 25(8) + 60 = 200 + 60 = 260$.

15 The ball goes up, reaches a peak, then comes down. Gravity causes the rate of change to vary, so the path is curved (nonlinear).

16 A $180°$ rotation uses $(x, y) \to (-x, -y)$. So $(0, 4) \to (0, -4)$.

17 In $\triangle ABC$: $\angle C = 180 - 55 - 80 = 45°$. Since $\angle C$ corresponds to $\angle F$ in congruent triangles, $\angle F = 45°$.

18 Reflecting over the y-axis negates x and keeps y: $(-4, 2) \to (4, 2)$.

19 All squares have four $90°$ angles. Any two squares can be mapped by a dilation since the ratio of their sides is constant. So all squares are similar.

20 Corresponding angles are in the same position (e.g., both upper-left) at each intersection of the transversal with the parallel lines.

21 $c^2 = 10^2 + 24^2 = 100 + 576 = 676$, so $c = \sqrt{676} = 26$.

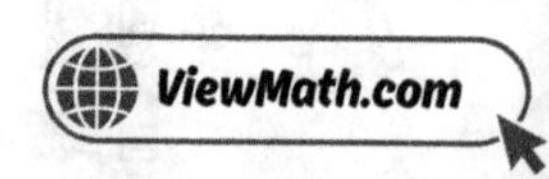

22. $d = \sqrt{(4-(-2))^2 + (9-1)^2} = \sqrt{6^2 + 8^2} = \sqrt{36 + 64} = \sqrt{100} = 10.$

23. $V = \frac{1}{3}(3.14)(64)(24) = \frac{1}{3}(4{,}823.04) \approx 1{,}607.7\ in^3.$

24. Complementary angles add to $90°$: $90 - 37 = 53°.$

25. Each interior angle $= \frac{(n-2)\times 180}{n} = 140.$ So $(n-2) \times 180 = 140n$, $180n - 360 = 140n$, $40n = 360$, $n = 9.$

26. The independent (explanatory) variable goes on the x-axis; the dependent (response) variable goes on the y-axis.

27. $m = \frac{4-10}{8-2} = \frac{-6}{6} = -1.$ $10 = -1(2) + b$, so $b = 12.$

28. $35 = 4x + 7$, $4x = 28$, $x = 7.$

29. Add rows and columns for totals. Divide each cell by 60 for relative frequency.

30. $P(A \text{ and } B) = 0.6 \times 0.4 = 0.24.$

☑ Practice Test 6 — Answer Key

 Rational    A    A    C    A    B    C    B    C

 C    A    A    Linear    B    B    C   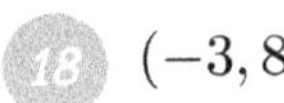 5 cm   18 $(-3, 8)$

 B    40    B    $\sqrt{13}$  B    C    B    B    A   28 B

 B   30 C

💡 Time to Learn! 💡

Review the explanations below, **especially for the questions you missed**.

Understanding why each answer is correct builds stronger problem-solving skills.

Tip: Circle any questions you got wrong, then read their explanation carefully.

📖 Practice Test 6 — Detailed Explanations

1. $\frac{5}{6}$ is a fraction of two integers with a nonzero denominator, so it is rational. Its decimal is $0.8\overline{3}$, which repeats.

2. $6^2 = 36$ and $7^2 = 49$. Since $36 < 45 < 49$, we have $6 < \sqrt{45} < 7$. The student is correct.

3. $60 \times 0.20 = 12$ discount, so the price is $60 - 12 = \$48$. Since $\$48 < \$50$, the discount is the better deal.

4. Any nonzero number raised to the zero power equals 1. So $(-2)^0 = 1$.

5. $Edge = \sqrt[3]{343} = 7$ in., since $7^3 = 343$.

6. Same exponent, so subtract the coefficients: $7.5 - 2.5 = 5$. Answer: 5×10^8.

7. $k = \frac{18}{6} = 3$, so the equation is $y = 3x$.

8. $\frac{270-120}{5-2} = \frac{150}{3} = 50$ mph.

9. $2x + 10 = 3x - 1$. Subtract $2x$: $10 = x - 1$. Add 1: $x = 11$.

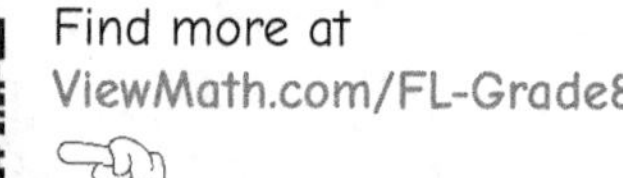
Find more at
ViewMath.com/FL-Grade8

10 $n + d = 15$ and $5n + 10d = 110$. From first: $n = 15 - d$. Substitute: $5(15 - d) + 10d = 110$, so $75 + 5d = 110$, $5d = 35$, $d = 7$.

11 $f(10) = 7 - 10 = -3$.

12 Function A is steeper — it rises faster. Its slope is 2 compared to Function B's slope of 1.

13 A constant increase of 6 means the rate of change is constant, which is the defining property of a linear function.

14 Starting height $b = 12$. It loses 2 inches per hour, so $m = -2$. The function is $y = -2x + 12$.

15 A steeper upward slope means faster increase. The morning section is steeper than the afternoon section, so the temperature rose faster in the morning.

16 Reflections are rigid transformations that preserve side lengths, angles, area, and perimeter.

17 $P = 2l + 2w$. So $28 = 2(9) + 2w$, giving $28 = 18 + 2w$, $2w = 10$, $w = 5$ cm.

18 Subtract the translation: $a = 2 - 5 = -3$, $b = 7 - (-1) = 8$. So $(a, b) = (-3, 8)$.

19 Similar figures have the same shape, which means all corresponding angles are equal. Side lengths, perimeter, and area may differ.

20 $90 + x + (x + 10) = 180$. Simplify: $2x + 100 = 180$, $2x = 80$, $x = 40$.

21 $c^2 = 1^2 + 1^2 = 2$, so $c = \sqrt{2}$.

22 $d = \sqrt{2^2 + 3^2} = \sqrt{4 + 9} = \sqrt{13}$.

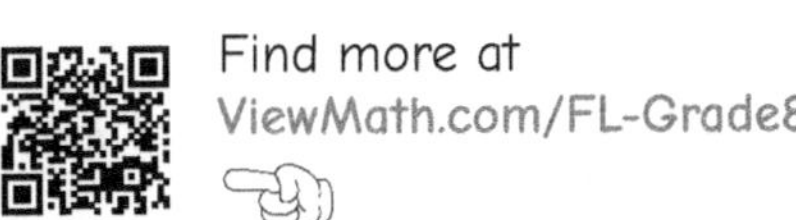

Find more at
ViewMath.com/FL-Grade8

23. $V = \frac{1}{3}\pi r^2 h = \frac{1}{3}\pi(9)(12) = \frac{108\pi}{3} = 36\pi \ m^3.$

24. $25 + 65 = 90°$. Complementary angles add to 90°.

25. An n-sided polygon divides into $n - 2$ triangles. $6 - 2 = 4$.

26. As x increases, y increases. The dots go up-right, indicating a positive association.

27. $m = \frac{8-3}{5-1} = \frac{5}{4}.$

28. Extrapolation means using the model to predict values outside the data range, which can be unreliable.

29. The 20 is the count for a specific combination of two variables (7th grade AND soccer), making it a joint frequency.

30. An impossible event has probability 0 because it can never occur.

✅ Practice Test 7 — Answer Key

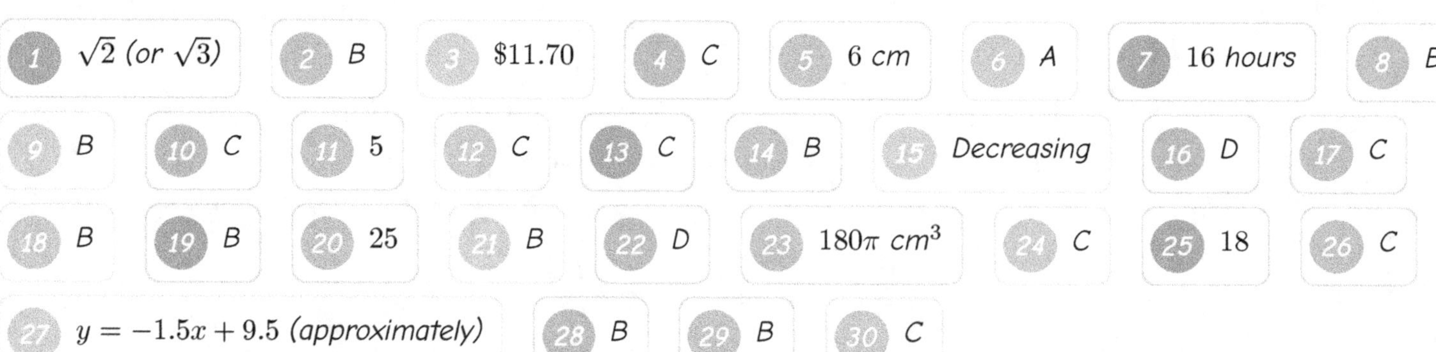

#	Answer		#	Answer		#	Answer
1	$\sqrt{2}$ (or $\sqrt{3}$)		2	B		3	$11.70
4	C		5	6 cm		6	A
7	16 hours		8	B		9	B
10	C		11	5		12	C
13	C		14	B		15	Decreasing
16	D		17	C		18	B
19	B		20	25		21	B
22	D		23	$180\pi \ cm^3$		24	C
25	18		26	C		27	$y = -1.5x + 9.5$ (approximately)
28	B		29	B		30	C

Find more at
ViewMath.com/FL-Grade8

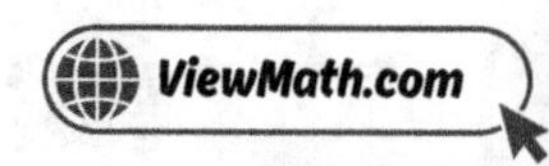

💡 **Time to Learn!** 💡

Review the explanations below, **especially for the questions you missed**.

Understanding why each answer is correct builds stronger problem-solving skills.

Tip: Circle any questions you got wrong, then read their explanation carefully.

📖 Practice Test 7 — Detailed Explanations

1 $\sqrt{2} \approx 1.414$ is between 1 and 2 and is irrational because 2 is not a perfect square. $\sqrt{3} \approx 1.732$ also works.

2 $\sqrt{5} \approx 2.236$ and $\frac{5}{2} = 2.5$. From least to greatest: $2 < 2.236 < 2.5$.

3 Tip $= 65 \times 0.18 = 11.70$. The tip is \$11.70.

4 Numerator: $9^3 \cdot 9^2 = 9^5$. Then $\frac{9^5}{9^4} = 9^{5-4} = 9^1 = 9$.

5 Edge $= \sqrt[3]{216} = 6$ cm, since $6 \times 6 \times 6 = 216$.

6 Multiply the coefficients: $3 \times 2 = 6$. Add the exponents: $10^{4+5} = 10^9$. So $(3 \times 10^4)(2 \times 10^5) = 6 \times 10^9$.

7 Rate $= \frac{6}{3} = 2$ hours per room. For 8 rooms: $2 \times 8 = 16$ hours.

8 $m = \frac{5-3}{5-1} = \frac{2}{4} = \frac{1}{2}$.

9 Subtract 7: $3x = 15$. Divide by 3: $x = 5$.

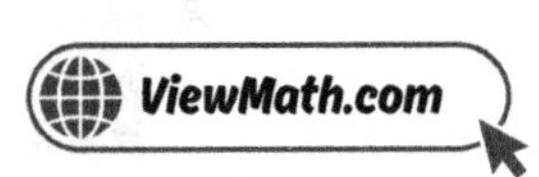

10 $2l + 2w = 34$ and $l = w + 5$. Substitute: $2(w + 5) + 2w = 34$, so $4w + 10 = 34$, $4w = 24$, $w = 6$.

11 $f(6) = \frac{6+4}{2} = \frac{10}{2} = 5$.

12 $\frac{9-1}{2-0} = \frac{8}{2} = 4$. The rate of change is 4.

13 $A = s^2$ has the variable raised to the 2nd power. For a function to be linear, the variable must have an exponent of 1.

14   Slope: $\frac{20-8}{3-0} = \frac{12}{3} = 4$. y-intercept: 8. So $y = 4x + 8$.

15   A graph that goes downward means the output (gas) is getting smaller over time.

16   A reflection mirrors (flips) a figure across a line of reflection.

17   Two figures are congruent if one can be mapped onto the other using translations, reflections, and/or rotations — all rigid transformations.

18   Divide the image coordinates by the scale factor. $(9 \div 3, \ -15 \div 3) = (3, -5)$.

19   Any circle can be mapped onto any other circle by a translation (to align centers) followed by a dilation (to match radii).

20 $(2x + 5) + 3x + (x + 25) = 180$. Combine: $6x + 30 = 180$, $6x = 150$, $x = 25$.

21 $a^2 = 17^2 - 8^2 = 289 - 64 = 225$, so $a = 15$. (Note: $\sqrt{225} = 15$, so B and D are the same value, but B is simplified.)

22 $d = \sqrt{(4-1)^2 + (5-1)^2} = \sqrt{9 + 16} = \sqrt{25} = 5$.

Find more at
ViewMath.com/FL-Grade8

23. $V = \pi(6^2)(5) = \pi(36)(5) = 180\pi \text{ cm}^3.$

24. y and $75°$ are vertical angles (across from each other), so $y = 75°$.

25. $\frac{(n-2) \times 180}{n} = 160.$ $180n - 360 = 160n.$ $20n = 360.$ $n = 18.$

26. A strong positive trend corresponds to an r-value close to 1. $r = 0.9$ indicates a strong positive association.

27. Using $(1, 8)$ and $(5, 2)$: slope $= \frac{2-8}{5-1} = \frac{-6}{4} = -1.5.$ $8 = -1.5(1) + b,$ $b = 9.5.$ Line: $y = -1.5x + 9.5.$

28. In $y = mx + b$, the slope m represents the rate of change: how much y changes for each 1-unit increase in x.

29. Of 42 pizza lovers, 24 are boys. $\frac{24}{42} \approx 0.571.$

30. There are 13 hearts in 52 cards: $P = \frac{13}{52} = \frac{1}{4}.$

✔ Practice Test 8 — Answer Key

1 C	**2** A: ≈ 4.2, B: ≈ 6.3, C: ≈ 8.5	**3** \$42
4 B	**5** -5	**6** C
7 A	**8** C	

9 C **10** 16 and 20 **11** C **12** C **13** $y = x^2$

14 $y = 12x + 50$; $m = 12$ is the hourly wage, $b = 50$ is the base pay.

15 Increasing, decreasing, increasing.

16 $(-a, -b)$ **17** $55°$ **18** C **19** $\frac{3}{2}$ **20** C **21** 13 ft **22** B **23** B **24** 10

25 $95°$ **26** C **27** Approximately 2 to 2.25 **28** C

Find more at
ViewMath.com/FL-Grade8

 **29** Boys: $\frac{20}{30} \approx 66.7\%$. Girls: $\frac{12}{20} = 60\%$. The percentages are close, so association is weak or none. **30** B 

💡 Time to Learn! 💡

Review the explanations below, **especially for the questions you missed.**

Understanding why each answer is correct builds stronger problem-solving skills.

Tip: Circle any questions you got wrong, then read their explanation carefully.

📖 Practice Test 8 — Detailed Explanations

1 Every integer n can be written as $\frac{n}{1}$, making it rational. Not every square root is irrational (e.g., $\sqrt{4} = 2$).

2 Side $= \sqrt{Area}$. $\sqrt{18} \approx 4.24 \approx 4.2$. $\sqrt{40} \approx 6.32 \approx 6.3$. $\sqrt{72} \approx 8.49 \approx 8.5$.

3 Markup $= 24 \times 0.75 = 18$. Selling price $= 24 + 18 = \$42$.

4 By the negative exponent rule, $\frac{1}{a^n} = a^{-n}$, so $\frac{1}{2^4} = 2^{-4}$.

5 $(-5)^3 = -125$, so $\sqrt[3]{-125} = -5$.

6 Planet A is 1.5×10^8 km away. Ten times that is 1.5×10^9 km, which matches Planet D.

7 Hose A: 15 gal/min. Hose B: $\frac{100}{8} = 12.5$ gal/min. Hose A is faster.

8 A slope of 0 means there is no rise — the line is horizontal.

 Find more at
ViewMath.com/FL-Grade8

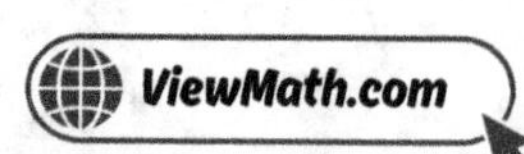 ViewMath.com

9 A true statement like $5 = 5$ means every value of x satisfies the equation.

10 $x = y + 4$ and $x + y = 36$. Substitute: $(y + 4) + y = 36$, $2y = 32$, $y = 16$. Then $x = 20$.

11 Find the column where $f(x) = 15$. That column has $x = 4$.

12 Function R: when $x = 0$, $y = 1$. Function S: $b = 1$. Both start at 1, so the initial values are equal.

13 Any equation where x has an exponent other than 1 (such as x^2, x^3, $\frac{1}{x}$, $\sqrt{x}$) is nonlinear.

14 Slope: $62 - 50 = 12$ (hourly rate). y-intercept: 50 (base pay before any hours). Equation: $y = 12x + 50$.

15 Filling: the water level rises. Drinking: the level drops. Refilling: the level rises again.

16 Reflect over x-axis: $(a, b) \to (a, -b)$. Then reflect over y-axis: $(a, -b) \to (-a, -b)$. This is the same as a $180°$ rotation.

17 $\angle R = 180 - 72 - 53 = 55°$. Since $\angle R$ corresponds to $\angle U$, $\angle U = 55°$.

18 The rule $(x, y) \to (-y, x)$ is a $90°$ counterclockwise rotation around the origin.

19 $\frac{9}{6} = \frac{3}{2}$, $\frac{12}{8} = \frac{3}{2}$, $\frac{15}{10} = \frac{3}{2}$. The scale factor is $\frac{3}{2}$.

20 By the Exterior Angle Theorem, y equals the sum of the two non-adjacent interior angles: $y = 55 + 75 = 130°$.

21 $d = \sqrt{12^2 + 5^2} = \sqrt{144 + 25} = \sqrt{169} = 13$ ft.

Find more at
ViewMath.com/FL-Grade8

22 The distance formula uses the horizontal and vertical distances as legs of a right triangle and applies $a^2 + b^2 = c^2$.

23 $V = \frac{4}{3}\pi r^3 = \frac{4}{3}(3.14)(27) = \frac{4}{3}(84.78) = 113.04 \ cm^3$.

24 $4x + 8 = 6x - 12$, so $20 = 2x$ and $x = 10$.

25 Sum $= 720°$. Sixth angle $= 720 - 5(125) = 720 - 625 = 95°$.

26 When data points group closely together, it is called a cluster.

27 Using endpoints: $m \approx \frac{11-2}{5-1} = \frac{9}{4} = 2.25$. A slope around 2 is a reasonable estimate.

28 For each 1-unit increase in x, y increases by the slope (6). For 3 units: $6 \times 3 = 18$.

29 Compare the conditional relative frequencies. Since 66.7% and 60% are similar, the preference doesn't change much by gender.

30 $P(not\ event) = 1 - P(event) = 1 - \frac{3}{4} = \frac{1}{4}$.

☑ Practice Test 9 — Answer Key

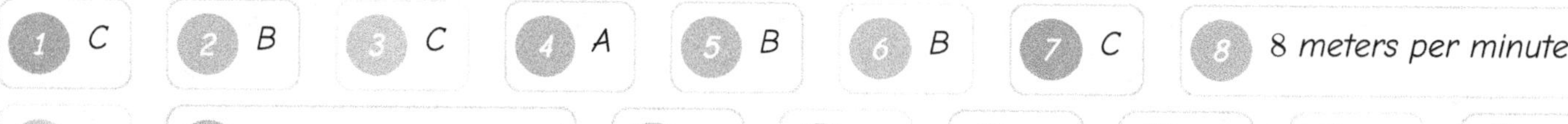

1	C	2	B	3	C	4	A	5	B	6	B	7	C	8	8 meters per minute
9	B	10	7 T-shirts and 8 hats	11	C	12	A	13	D	14	B	15	C	16	B
17	A	18	B	19	B	20	A	21	C	22	B	23	B	24	A	25	B	26	A
27	A	28	B	29	C

Find more at
ViewMath.com/FL-Grade8

Independent: flipping a coin twice (first flip doesn't affect second). Dependent: drawing cards without replacement (first

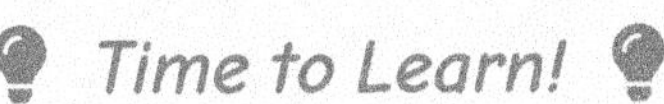

💡 Time to Learn! 💡

Review the explanations below, **especially for the questions you missed**.

Understanding why each answer is correct builds stronger problem-solving skills.

Tip: Circle any questions you got wrong, then read their explanation carefully.

📖 Practice Test 9 — Detailed Explanations

1. $1.41421356\ldots$ is the decimal expansion of $\sqrt{2}$, which is non-repeating and non-terminating, indicating an irrational number.

2. $3^2 = 9$ and $4^2 = 16$. Since $9 < 10 < 16$, we have $3 < \sqrt{10} < 4$.

3. Discount $= 120 \times 0.25 = 30$. Sale price $= 120 - 30 = \$90$.

4. $3^{-2} = \frac{1}{3^2} = \frac{1}{9}$.

5. $7^2 = 49$ and $8^2 = 64$. Since $49 < 50 < 64$, we have $7 < \sqrt{50} < 8$.

6. $(2.4 \times 10^3)(5 \times 10^2) = 12 \times 10^5 = 1.2 \times 10^6$.

7. $k = \frac{12}{8} = 1.5$. At $x = 20$: $y = 1.5 \times 20 = 30$.

8. $m = \frac{40-0}{5-0} = \frac{40}{5} = 8$ meters per minute.

Find more at
ViewMath.com/FL-Grade8

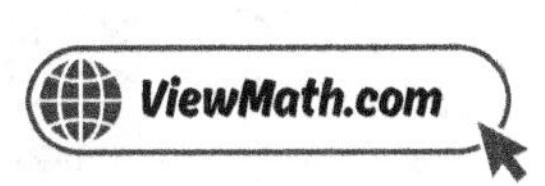

9. $4x - 12 = 2x + 2$. Subtract $2x$: $2x - 12 = 2$. Add 12: $2x = 14$, so $x = 7$.

10. $t + h = 15$ and $12t + 8h = 148$. From first: $h = 15 - t$. Substitute: $12t + 8(15 - t) = 148$, $4t + 120 = 148$, $4t = 28$, $t = 7$. Then $h = 8$.

11. $f(-3) = -(-3) + 8 = 3 + 8 = 11$.

12. The rate of change is the slope. Function A has slope 5 and Function B has slope 2. Since $5 > 2$, Function A has the greater rate of change.

13. $y = \sqrt{x}$ involves a square root of the variable, so it is nonlinear. The others all fit $y = mx + b$.

14. Slope: $\frac{11-5}{2-0} = \frac{6}{2} = 3$. The y-intercept from $(0,5)$ is $b = 5$. So $y = 3x + 5$.

15. A function can decrease while staying positive (e.g., dropping from 10 to 5). "Decreasing" means the values get smaller, not that they are negative.

16. Translations preserve all side lengths. The longest side is still 8 cm.

17. Both translation and rotation are rigid transformations. The image is congruent to the original regardless of the shape.

18. The $90°$ CCW rule is $(x, y) \to (-y, x)$. So $(-3, 5) \to (-5, -3)$.

19. When $k = 1$, every point stays in the same place. The image is congruent (identical in size and shape) to the original.

20. $\angle C = 180 - 45 - 85 = 50°$. All three angles are less than $90°$, so the triangle is acute.

21. $9^2 + 40^2 = 81 + 1600 = 1681 = 41^2$. The converse of the Pythagorean Theorem confirms it's a right triangle.

Find more at
ViewMath.com/FL-Grade8

ViewMath.com

22 $d = \sqrt{3^2 + 4^2} = \sqrt{9 + 16} = \sqrt{25} = 5$.

23 $80\pi = \frac{1}{3}\pi(16)h$. Divide both sides by π: $80 = \frac{16h}{3}$. Multiply by 3: $240 = 16h$. Divide: $h = 15$ cm.

24 Supplementary angles add to $180°$: $180 - 125 = 55°$.

25 Sum of pentagon angles $= 540°$. $x = 540 - 110 - 105 - 115 - 100 = 110°$.

26 Clusters are groups of points that bunch together, while outliers are isolated far from the trend. Both can appear in the same scatter plot.

27 The line goes from about $(0.5, 6.5)$ to $(7.5, 1.5)$. Slope $\approx \frac{1.5 - 6.5}{7.5 - 0.5} = \frac{-5}{7} \approx -0.71$.

28 $y = 5(8) + 20 = 40 + 20 = 60$ dollars.

29 Two-way tables deal with categorical data (counts and proportions). Mean scores involve numerical data, not categorical.

30 Independent events don't affect each other's probabilities. Dependent events do — the first outcome changes the conditions for the second.

✓ Practice Test 10 — Answer Key

 B    B    A    C    C   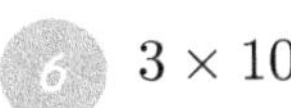 3×10^7  12 cups    3    D

 B    9    B    Differences are all 4; the function is linear.   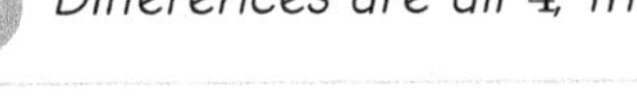 B    B

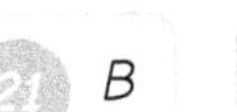 $(-5, -3)$  C   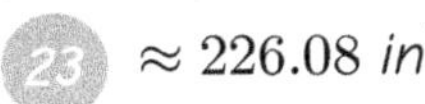 $(-5, -2)$  A    B    B   22 B   23 ≈ 226.08 in^3

 **24** B  **25** C 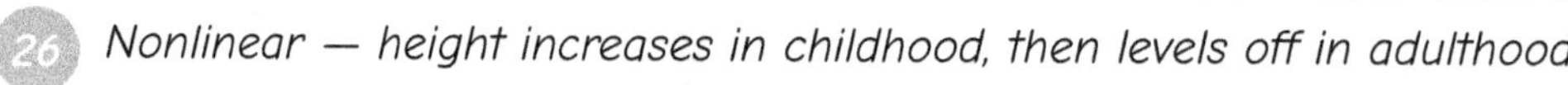 **26** Nonlinear — height increases in childhood, then levels off in adulthood.  **27** C

 **28** C  **29** A  **30** A

💡 Time to Learn! 💡

Review the explanations below, **especially for the questions you missed.**

Understanding why each answer is correct builds stronger problem-solving skills.

Tip: Circle any questions you got wrong, then read their explanation carefully.

📖 Practice Test 10 — Detailed Explanations

1 5 is not a perfect square, so $\sqrt{5}$ is irrational and belongs in the Irrational region.

2 $\sqrt{6} \approx 2.449$. Point B is located at approximately 2.45 on the number line, which is the closest to $\sqrt{6}$.

3 $Tax = 80 \times 0.07 = 5.60$. $Total = 80 + 5.60 = \$85.60$.

4 A negative exponent flips the fraction: $\left(\frac{3}{4}\right)^{-2} = \left(\frac{4}{3}\right)^2 = \frac{16}{9}$.

5 $\sqrt{2}$ cannot be expressed as a fraction of two integers, so it is irrational.

6 $\frac{7.2 \times 10^{12}}{2.4 \times 10^5} = \frac{7.2}{2.4} \times 10^{12-5} = 3 \times 10^7$.

7 $k = \frac{2}{5}$ cups per cookie. For 30 cookies: $\frac{2}{5} \times 30 = 12$ cups.

8 $m = \frac{11-(-1)}{7-3} = \frac{12}{4} = 3$.

Find more at
ViewMath.com/FL-Grade8

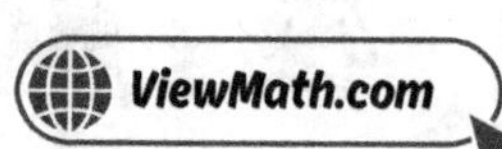

9. Subtracting $6x + 4$ from both sides gives $0 = 0$, which is always true. Every value of x is a solution.

10. $p + n = 8$ and $2p + 5n = 25$. From first: $p = 8 - n$. Substitute: $2(8 - n) + 5n = 25$, so $16 + 3n = 25$, $3n = 9$, $n = 3$.

11. From the graph, $f(1) = 3$ and $f(4) = 6$. So $f(1) + f(4) = 3 + 6 = 9$.

12. Set $2x + 100 = 8x + 10$. Subtract $2x$: $100 = 6x + 10$. Subtract 10: $90 = 6x$. Divide: $x = 15$.

13. $1 - (-3) = 4$, $5 - 1 = 4$, $9 - 5 = 4$, $13 - 9 = 4$. The constant difference of 4 means the rate of change is constant, so the function is linear.

14. Starting amount $b = 120$. Spending \$15/day means losing money, so $m = -15$. The function is $y = -15x + 120$.

15. A steeper distance-time graph means more distance covered per unit of time — that means greater speed.

16. Reflecting over the y-axis flips the sign of the x-coordinate: $(5, -3) \rightarrow (-5, -3)$.

17. Two circles with the same radius are always congruent — one can be translated to overlap the other.

18. $90°$ CCW: $(x, y) \rightarrow (-y, x)$. So $(-2, 5) \rightarrow (-5, -2)$.

19. Perimeter scales by the same factor. $36 \times \frac{1}{3} = 12$ cm.

20. Co-interior angles are supplementary: $180 - 135 = 45°$.

21. $c^2 = 6^2 + 8^2 = 36 + 64 = 100$, so $c = 10$.

Find more at
ViewMath.com/FL-Grade8

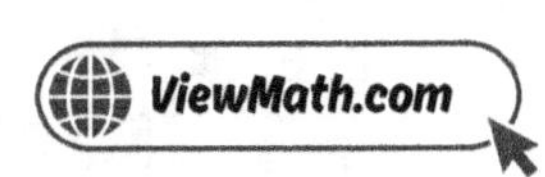

22 The x-coordinates are the same, so the distance is the vertical difference: $|7 - (-5)| = 12$.

23 $V = 3.14 \times 9 \times 8 = 226.08 \ in^3$.

24 The two angles are complementary: $(2x + 4) + 36 = 90$. So $2x + 40 = 90$, $2x = 50$, $x = 25$.

25 $(n - 2) \times 180 = 900$. $n - 2 = 5$. $n = 7$ (heptagon).

26 Height rises quickly in youth, slows, and plateaus in adulthood. This is a nonlinear pattern.

27 A good line of best fit has roughly half the points above and half below.

28 $y = 2(4) + 1 = 9$.

29 $\frac{24}{80} = 0.30$.

30 With replacement: $P(red) = \frac{5}{8}$, $P(blue) = \frac{3}{8}$. $P = \frac{5}{8} \times \frac{3}{8} = \frac{15}{64}$.

Well done checking your answers!

Keep practicing to strengthen your skills.

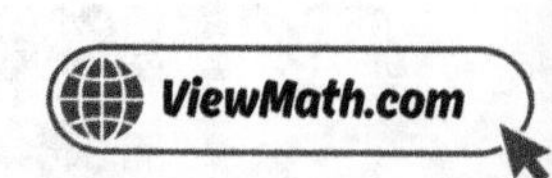